MIX
Papier aus verantwortungsvollen Quellen
Paper from responsible sources
FSC® C105338

**Ashish Bohre
Kalpana Awasthi**

Immobolization of radioactive waste in ceramic based hosts

Radioactive waste Immobolization

Anchor Academic
Publishing

Bohre, Ashish, Awasthi, Kalpana: Immobolization of radioactive waste in ceramic
based hosts : Radioactive waste Immobolization. Hamburg, Anchor Academic
Publishing 2013

Buch-ISBN: 978-3-95489-169-6
PDF-eBook-ISBN: 978-3-95489-669-1
Druck/Herstellung: Anchor Academic Publishing, Hamburg, 2013

Bibliografische Information der Deutschen Nationalbibliothek:
Die Deutsche Nationalbibliothek verzeichnet diese Publikation in der Deutschen
Nationalbibliografie; detaillierte bibliografische Daten sind im Internet über
http://dnb.d-nb.de abrufbar.

Bibliographical Information of the German National Library:
The German National Library lists this publication in the German National Bibliography.
Detailed bibliographic data can be found at: http://dnb.d-nb.de

All rights reserved. This publication may not be reproduced, stored in a retrieval system
or transmitted, in any form or by any means, electronic, mechanical, photocopying,
recording or otherwise, without the prior permission of the publishers.

Das Werk einschließlich aller seiner Teile ist urheberrechtlich geschützt. Jede Verwertung
außerhalb der Grenzen des Urheberrechtsgesetzes ist ohne Zustimmung des Verlages
unzulässig und strafbar. Dies gilt insbesondere für Vervielfältigungen, Übersetzungen,
Mikroverfilmungen und die Einspeicherung und Bearbeitung in elektronischen Systemen.

Die Wiedergabe von Gebrauchsnamen, Handelsnamen, Warenbezeichnungen usw. in
diesem Werk berechtigt auch ohne besondere Kennzeichnung nicht zu der Annahme,
dass solche Namen im Sinne der Warenzeichen- und Markenschutz-Gesetzgebung als frei
zu betrachten wären und daher von jedermann benutzt werden dürften.

Die Informationen in diesem Werk wurden mit Sorgfalt erarbeitet. Dennoch können
Fehler nicht vollständig ausgeschlossen werden und die Diplomica Verlag GmbH, die
Autoren oder Übersetzer übernehmen keine juristische Verantwortung oder irgendeine
Haftung für evtl. verbliebene fehlerhafte Angaben und deren Folgen.

Alle Rechte vorbehalten

© Anchor Academic Publishing, Imprint der Diplomica Verlag GmbH
Hermannstal 119k, 22119 Hamburg
http://www.diplomica-verlag.de, Hamburg 2013
Printed in Germany

Table of Contents

Chapter 1 1
Ceramic Materials and Investigation of Literatures 1

   1.1   Different Ceramic materials 1
   1.2   Structures of Ceramic materials 2
   1.3   Ceramics Properties 3
   1.4   Conventional ceramics 3
   1.5   Modern ceramics and their applications 4
   1.6   Future Scope of the present work 12
   1.7   Investigation of relevant literature 14

      Metal Substituted sodium zirconium phosphate (NZP) 14
      Metal Substituted calcium zirconium phosphate (CZP) 16
      Metal Substituted perovskites 19
      Substituted calcium titanate 20
      References 22

Chapter 2 28
Crystal Structure Refinement and instrumentation 28

   2.1.   Powder Diffraction method 28
   2.2.   Determination of crystal structure 30
       2.2.1..   Conventional approaches 31
       2.2.2.   Straight space approaches 31
   2.3.   Refinement of crystal structure 32
       2.3.1.   Rietveld Refinement Method 32
       2.3.2.   Rf factors 36
   2.4.   Application of Different software for solving crystal structure 38
       2.4.1.   Indexing programs 38
       2.4.2.   General Structure analysis system software 40
   2.5.   Scanning electron microscopy 43
   2.6.   Energy Dispersive X-ray microanalysis 44
   2.7.   Impedance Spectroscopy 46
   2.8.   FTIR Spectroscopy 49
      References 51

Chapter 3 54
Investigational Procedures 54
   3.1   Characterization and synthesis of materials 54
   3.2   Characterization of ceramic materials 57
      3.2.1   Analysis of phases using rietveld refinement 57
      3.2.2   Morphology and elemental analysis using SEM/EDAX 57
   3.3   Electrical properties Study 58
      3.3.1   Dielectric measurements 58
   3.4   FTIR Spectroscopy 59
   3.5   Experimental data 59

Chapter 4 141
Outcomes of the Research Work 141
   4.1   Metal Substituted Sodium Zirconium Phosphates (NZP) 141
      4.1.1 Metal Strontium substituted sodium zirconium phosphate: 141
      4.1.2   Cesium / Strontium substituted sodium zirconium phosphates: 154
   4.2   Metal Substituted calcium zirconium phosphates: 163
      4.2.1   Molybdenum Substituted calcium zirconium phosphates: 163
   4.3   Perovskites 171
      4.3.1   Calcium Samarium Titanate: $Ca_{1-x}Sm_xTiO_3$ (where x = 0.1-0.5) 171
      References 187

Conclusion 191
Summary 194

Chapter 1

Ceramic Materials and Investigation of Literatures

1.1 Different Ceramic materials

Ceramic materials are inorganic, non-metallic solids, which consist of an aggregate of randomly oriented crystallites bonded together by ionic bonds and have covalent character. In contrast, the Anglo-Saxon term "ceramics" also often includes glass, enamel, glass-ceramic, and inorganic cementitious materials (cement, plaster and lime). Hence ceramic materials are polycrystalline materials that acquire their mechanical strength through various sintering processes. Ceramics are good thermal and electric insulators; more stable having high melting point, high chemical resistance and high compressive strength.

"Ceramic materials find application in daily life e.g. electronic components, environment sensors, gas ignitors, ultrasonic cleaner and intrusion alarm etc. There are many combinations of metallic and non-metallic atoms that can combine to form ceramic components and also several structural arrangements are usually possible for each combination of atoms. This led scientists to invent many new ceramic materials to meet increasing requirements and demands in various application areas. Advanced furnaces and heat engines played important roles in the success of the industrial revolution, while ceramic materials were essential for thermal insulation of various types of furnaces and engines. As higher and higher frequencies and voltages were used, the demand on ceramic dielectrics became more stringent. Also, new specifications for the magnetic and optical properties of ceramics were developed as a part of the new electronic and electro–optical technology revolution".

Advanced ceramics offer numerous enhancements in performance, durability, reliability, hardness, high mechanical strength at high temperature, stiffness, low density, optical conductivity, electrical insulation and conductivity, thermal insulation and conductivity, radiation resistance, and so on. Ceramic technologies have been widely used for aircraft and aerospace applications, wear-resistant parts, bioceramics, cutting tools, advanced optics, superconductivity, nuclear reactors, etc.

1.2 Structures of Ceramic materials

The structure of ceramic materials is dictated by the type of atoms present, the type of bonding between the atoms, and the way the atoms are packed together. The atoms in ceramic materials are held together by a chemical bond and the two most common chemical bonds for ceramic materials are covalent and ionic. For metals, the chemical bond is called metallic bond. The bonding of atoms together is much stronger in covalent and ionic than in metallic bonding. That is why, generally speaking, metals are ductile and ceramics are brittle. The primary bonding mechanism in ceramics is ionic bonding. Ionically bonded materials are crystalline in nature and have both a high electrical resistance and a high relative dielectric constant. A degree of covalent bonding as a result of sharing of electrons in the outer shell may also be present, particularly in some of the silicon and carbon-based ceramics. Both types of bonds are very strong, resulting in ceramics that have a high melting point, are very stable chemically, and are not attacked by ordinary solvents and most acids.

1.3 Ceramics Properties

Ceramics possess a wide range of physical and chemical properties making them useful for virtually every aspect of modern life. Most of the physical and chemical properties of ceramics originate from their crystal structure, chemical composition and defect chemistry.

The most important properties in ceramic materials are:

- High Strength
- High Fracture Toughness
- High Hardness
- Excellent Wear Resistance
- Good Frictional behavior
- Resistant to plastic deformation
- Resistant to high temperatures
- Good corrosion resistance
- Low electrical conductivity

- Refractory
- Thermal & Electrical insulators
- Excellent Surface Finish
- Nonmagnetic
- Oxidation resistant
- Low thermal conductivity
- Modulus of Elasticity
- Brittle
- Anti-Static

1.4 Conventional ceramics

"Traditional ceramics are derived from common, naturally occurring raw materials such as clay minerals and quartz sand. Traditional ceramic objects are almost as old as the human race. Naturally occurring abrasives were undoubtedly used to sharpen primitive wood and stone tools, and fragments of useful clay vessels have been found dating from the Neolithic Period, some 10,000 years ago. Not long after the first crude clay vessels were made, people learned how to make them stronger, harder, and less permeable to fluids by burning. These advances were followed by structural clay products, including brick and tile. Clay-based bricks, strengthened and toughened with fibers such as straw, were among the earliest composite materials. Artistic uses of pottery also achieved a high degree of sophistication, especially in China, the Middle East, and the Americas.

History and ceramics are intertwined. Advances in civilization have always followed advances or innovations in materials. As archaeologists and anthropologists tell us, one of the first steps in human development was taken when early cultures learned to use natural materials, such as wood and rock, as tools and weapons. The next step began when they learned to use rocks to chip other rocks, such as chert and obsidian (volcanic glass), into more efficient tools and weapons. So important was this use of natural ceramic materials that the prehistoric time in which it occurred is now referred to as the Stone Age. But the Stone Age was just the beginning of the use of materials to improve our standard of life. Eventually, people learned to make pottery; to extract and use metals (the Bronze Age and Iron Age); to produce glass; and to make bricks, tiles, and cement. Much later materials made possible the Industrial Revolution, the harnessing of electricity, and the "horseless carriage." Within the last fifty years or so materials ushered in the age of electronics, the jet airplane, near-instantaneous worldwide communications, and the exploration of outer space".

1.5 Modern ceramics and their applications

Advanced ceramics known as engineering or technical ceramics refer to materials which exhibit superior mechanical properties, corrosion/oxidation resistance, and thermal, electrical, optical or magnetic properties.

The beginning of the advanced ceramics era has been said to have started approximately 50 years ago with the expanding use of chemically prepared powders. For example, the Bayer process for the production of alumina powders initially grew from spark plug production. While these powders would be considered relatively low grade by today's standards, they were more pure and offered more control over the composition, microstructure, and crystal structure over minerals-based ceramics.

Advanced ceramics are generally broken down into the following segments:
- Structural ceramics
- Electrical and electronic ceramics
- Ceramics coatings
- Bioceramics
- Nanoceramics

Structural ceramics

"Structural ceramics materials demonstrate enhanced mechanical properties under demanding conditions. Because they serve as structural members, often being subjected to mechanical loading, they are given the name structural ceramics. Ordinarily, for structural applications. ceramics tend to be expensive replacements for other materials, such as metals, polymers, and composites. For especially erosive, corrosive, or high-temperature environments, however, they may be the material of choice. This is because of the strong chemical bonding in ceramics. Chemical bonds make them exceptionally robust in demanding situations.

Advanced ceramics are characterized by

- High performance in terms of mechanical, electrical, chemical and thermal properties
- Predominantly inorganic-nonmetallic behavior
- High value-added properties through a sophisticated process technology
- Unique functional properties
- Their position at the beginning of the development cycle

Structural advanced ceramics are used for applications where a component of an engineering system is subjected to high mechanical, thermal or chemical loads. Typical structural ceramics are alumina, partially and fully stabilized zirconia, cordierite, silicon nitride, silicon carbide, boron nitride, titanium nitride, and titanium boride. Zirconia ceramics excel in high temperature applications such as thermal barrier coatings where high density, high bending strength, high thermal expansion and low thermal conductivity are required. Silicon carbide ceramics possess high thermal conductivity and high micro hardness, but are deficient in strength and thermal shock resistance. Hence, the area of application of such ceramics is in heating elements for furnaces operating up to 1600°C in ambient atmospheres. Silicon nitride on the other hand shows a high thermal shock resistance, sufficient strength and hardness, and intermediate values of density and thermal conductivity which suggest applications in automotive engines and heat exchangers, as well as for ceramic cutting tools for the high speed machining of hard metals. Lastly, alumina is hard and shows a rather high thermal expansion, but performs badly in terms of bending strength and thermal shock resistance".

Electrical and electronic ceramics

"Advanced electro ceramics have played a critical role in the development of new and modern technologies such as computers, telecommunications and aerospace and they will continue to play the leading role in the technology of the future. The materials cobalt-mangan oxides used as negative temperature coefficient thermistors, semi conducting doped barium titanate used for positive temperature coefficient thermistors with large applications in thermal protections and piezoelectric electro ceramics of PZT type, which find new application as ceramic transducers and sensors in biomedical and aerospace industries or as simply buzzers, filters, igniters, ultrasonic cleaners, towed array sonar's, medical imaging system and ink-jet printing heads. Electronic ceramics are integral components of numerous electronic devices including computers, wireless communication, automotive and industrial control systems, and digital switches. High performance refractories are used by manufacturers of ceramic capacitors, integrated circuit packages, and ferrites.

Demand for microwave dielectric ceramics used in telecommunications, as well as in cable communications using optical fibers, is increasing rapidly. Whereas, in the past microwave communications were used primarily for military purposes such as radar, weapon guidance systems and satellite communications, more recently microwaves have been utilized extensively in communications devices such as mobile radios and phones, and in satellite broadcasting.

The perovskite families of materials are important in many electro ceramic applications. Perovskite based ceramics are used in ferroelectric, pyroelectric, piezoelectric, superconductor, thermoelectric, dielectric, magnetic, linear and nonlinear electro-optic devices.

Ceramics coatings

Ceramic coatings are a special film layer used on hard materials to protect them from wear and tear. They play a big role in keeping the strength and quality of materials such as metals, plastic and many other products used in industrial and consumer applications. Ceramic coatings are typically made up of special compounds that possess stronger properties than regular coating. Compounds in ceramic coatings include carbides, borides, nitrides, and silicides that are hard enough to keep the material from accumulating residue and becoming prone to breakage while in use. The sputtering process is used for coating. This removes certain portions of the coating material, and deposits a thin and securely bonded film onto the surface. During the sputtering process, elements such as nitrogen and hydrocarbon are added to the ceramic coating. When added, these elements make the solution more flexible and elastic.

There are two types of ceramic coatings, single-layer and multi-layer. Single-layer coatings are usually applied on already strong materials that are not always exposed to damage, such as PVC plastic and non-metallic products. Multi-layer coatings are used on more delicate products such as fiberglass, metals and other items used for special laboratory or industrial purposes.

Among the properties in which ceramic coatings are known for include improved strength, hardness, elasticity, and oxidation resistance. Oxidation resistance is the most important feature, as it keeps the product not only from breaking or being damaged, but also from rusting and other effects of oxidation due to exposure to various airborne agents.

Ceramic coatings play a dominating role in a large number of applications. They can perform functional tasks, such as piezo-electrical layers, controllable dielectrics, coatings for the photovoltaic energy conversion or electrolyte-layers in high-temperature fuel cells, which are only part of the whole range of applications possible. Moreover, ceramic coatings very often serve to protect load-bearing components made of other materials, mainly metals. In such applications, they are used to increase hardness, scratch resistance or to improve tribiological properties in general. Ceramic coatings are being used as a barrier between dissimilar metals to reduce friction, which in turn reduces wear in internal engine components. The most common applications for ceramic coatings are on the exhaust system, intake manifolds, and exhaust headers.

Bioceramics

One of the most exciting areas of medical advancement during our generation has been the repair or complete replacement of damaged parts of our bodies. For a replacement to be successful, the material used must be able to perform the same function as the original body part and also must be compatible with the surrounding tissue. These sound like fairly simple requirements but actually are quite challenging, especially with regard to compatibility. Many materials are toxic to tissue. Most others that aren't toxic instead are attacked by the body and either destroyed or encapsulated in special fibrous tissue (similar to scar tissue) that our body builds up in an effort to isolate the foreign part that has been implanted. Some crystalline ceramics and glass, however, mimic the composition of bone (which is mostly ceramic) and have proved to be the most body-compatible materials found so far. These biocompatible ceramic materials are referred to as bioceramics.

Bioceramics are now successfully implanted in our bodies as solid parts, porous parts, and coatings. Those that actually bond with our bone and tissue are called bioactive. Bioglass and polycrystalline ceramics with similar composition to Bioglass, such as a ceramic called hydroxyapatite, are the most successful bioactive ceramics. Some of these compositions are actually slowly absorbed by our body as they are replaced with natural bone and tissue. Other ceramic materials, such as alumina and zirconia, don't bond to our bone and tissue but are neither toxic nor attacked by our body and rejected. These inert ceramics also have become important as implants, especially for the replacement of damaged or worn out joints.

Bioceramics can have structural functions as joint or tissue replacements, can be used as coatings to improve the biocompatibility of metal implants, and can function as resorbable lattices which provide temporary structures and a framework that is dissolved, replaced as the body rebuilds tissue. Some ceramics even feature drug-delivery capability.

Ceramics in a number of forms and compositions are currently in use or under consideration, with more in development. Alumina and zirconia are among the bioinert ceramics used for prosthetic devices. Bioactive glasses and machinable glass-ceramics are available under a number of trade names. Porous ceramics such as calcium phosphate-based materials are used for filling bone defects. The ability to control porosity and solubility of some ceramic materials offers the possibility of use as drug delivery systems. Glass microspheres have been employed as delivery systems for radioactive therapeutic agents, for example. The thermal and chemical stability of ceramics, high strength, wear resistance and durability all contribute to making ceramics good candidate materials for surgical implants.

Nanoceramics

Ceramics is one of the fields where nanoscience and nanotechnology have shown remarkable progress, producing a variety of advanced materials with unique properties and performance. Nanoceramics is a term used to refer to ceramic materials fabricated from ultra fine particles, i.e., less than 100 nm in diameter. In this field, a great deal of research has been accomplished in the last 20 years and has resulted in significant outcomes that are of great impact academically as well as industrially. Nanocrystalline ceramic materials exhibit the outstanding mechanical properties highly attractive for a wide range of applications. In particular, nanocrystalline ceramic bulk materials and coatings are often characterized by superstrength, superhardness and good wear resistance.

Nanostructure ceramics are attractive materials that find potential uses ranging from simple everyday applications like paints and pigments to sophisticated ones such as bioimaging, sensors, etc. The inability to economically synthesize nanoscale ceramic structures in a large scale and simultaneously achieve precise control of their size has restricted their real time application. Electro spinning is an efficient process that can fabricate Nanofibers on an industrial scale. During the last 5 years, there has been remarkable progress in applying this process to the fabrication of ceramic nanorods and nanofibers. Ceramic nanofibers are becoming useful and niche materials in several applications owing to their surface dependant and size dependant properties. Carbon nanotubes are molecular-scale tubes of graphitic carbon with outstanding properties. They are among the stiffest and strongest fibres known, and have remarkable electronic properties and many other unique characteristics. A ceramic material reinforced with carbon nanotubes has been made by materials scientists. The new material is far tougher than conventional ceramics, conducts electricity and can both conduct heat and act as a thermal barrier, depending on the orientation of the nanotubes. The researchers mixed powdered alumina (aluminum oxide) with 5 to 10 percent carbon nanotubes and a further 5 percent finely milled niobium. Carbon nanotubes are sheets of carbon atoms rolled up into tiny hollow cylinders. With diameters measured in nanometers, they have unusual structural and conducting properties".

Some other application of ceramics

- **Lasers**

"Ceramic laser technology has emerged as a promising candidate because of its numerous advantages over single-crystal lasers. First, ceramics can be produced in large volumes, which makes them attractive for high-power laser generation. Second, they can provide a gain medium for fiber lasers with high beam quality and can also be made into composite laser media with complicated structures that would otherwise be difficult to fabricate. Besides, ceramics can be heavily and homogeneously doped with laser-active ions. They can also be used to fabricate novel laser materials, such as sesquioxides, which cannot be produced by the conventional melt–growth process. In addition, single crystals with new structures can be fabricated from ceramics through sintering. This type of ceramic-derived single crystal has proved to have high resistance to laser damage and a long lifetime, and is very promising for high-power-density lasers. This novel laser gain medium cannot be produced by conventional single-crystal-growth technology and may offer new laser performance. Solid-state lasers are widely used in metal processing medical applications, such as eye surgery, red–green–blue (RGB) light sources in laser printers and projectors, environmental instrumentation measurements and optical transmission systems, and have demonstrated potential for future nuclear-fusion applications.

- **Ceramics for high temperature industrial process**

    Many of the ways ceramics are used in industrial processes are similar to those in metals processing: for furnace linings, high-temperature fixtures, burners, and heat exchangers. These parts are required in glass production, ceramics production, chemicals processing, petroleum refining, and even papermaking. Many industrial processes require a heat source at some stage. Some examples include paint drying and window shaping in the automotive industry, food processing, paper drying, preparation of cement and plaster, and disposal of hazardous materials. Ceramics are involved in several different ways. In some cases, the heat is produced by burning a fuel. The chamber in which the fuel is burned generally is lined with ceramics or made completely of ceramics.

- **The Space applications**

    Our journeys into the sky, and even into outer space, require the services of ceramic materials in many different ways, but especially for resistance to high temperature. Probably the most familiar use of ceramics for aerospace technology is in the Space Shuttle. The outer surface of the Space Shuttle is covered with ceramic tiles that protect the underlying aluminum structure and the astronauts from getting too hot during ascent into space and reentry into the Earth's atmosphere. The friction of the Shuttle traveling at high speed through the atmosphere produces temperatures up to about 2650°F about double the melting temperature of the aluminum alloys from which most of the Shuttle is built.

    Two types of carefully engineered ceramics protect the Shuttle. The tip of the nose and the front edges of the wings, which are exposed to the highest temperatures, are made from a ceramic composite of carbon fibers in a carbon matrix. This composite has extremely high temperature resistance but burns if it comes in contact with the oxygen in the air at high temperature. To protect against contact with oxygen, the carbon-carbon composite is first coated with a layer of silicon carbide and then over coated with a layer of silicon dioxide.

- **Nuclear Energy**

    Ceramics are necessary in the nuclear power industry as fuel pellets, control rods, high reliability seals and valves, and a special means of containing (encapsulating) and stabilizing radioactive wastes for long periods of time. The fuel pellets used in a conventional reactor are made mostly of uranium oxide ceramic. A pellet only about three-eighths of an inch in diameter and one-half an inch long contains as much energy as 1 ton of coal, 150 gallons of oil, or 22,500 cubic feet of natural gas. The fuel pellets for a nuclear reactor are enclosed in rods over 12 feet long. A typical reactor contains over 45,000 fuel rods and can produce enough electricity to meet the needs of 350,000 all-electric homes.

- **Ceramics and Renewable Energy Sources**

    The burning of oil and coal consumes natural resources that required millions of years to form and that cannot be replenished. We need to switch more and more emphasis to energy sources that can be replenished, referred to as *renewable energy sources*. Two renewable energy sources that have been used for many years are windmills and hydroelectric dams, to harvest energy from wind and water, respectively. Ceramics are important in both technologies. Modern windmill blades are fabricated from composites containing ceramic-fiber reinforcement. Dams generally are constructed from concrete. Both methods of harvesting energy also require electrical ceramics, especially electrical insulators. They also require machinery that rotates at high speed, which can benefit from new wear-resistant ceramics, especially the new silicon nitride ceramic bearings".

1.6 Future Scope of the present work

Ceramic materials have found numerous applications in science, technology and industry as mentioned earlier. One of the recent applications of titania and zirconia based ceramic precursors is in immobilization and solidification of radioactive isotopes in waste effluents coming out of nuclear establishments and power plants. Due to long term stability and integrity of the ceramic waste forms of high and intermediate level nuclear waste, several countries have now switched over from 'glass technology' to 'ceramic technology' of radwaste management. These and many more applications make ceramics materials an interesting area of research and engineering sciences.

It would, therefore, be desirable to study various routes of synthesis of titania and zirconia based ceramic materials followed by their physiochemical characterization and possible applications as industrial materials. There are several solid–state reactions, which takes place on verification of high-level nuclear waste. In order to develop and appreciate the "structure-property" relationship of various post vitrified ceramic phases, it has become logical and interesting to *simulate the solid- state reactions that might takes place in the process of conversion of nuclear effluents into the corresponding ceramic waste forms.* The primary objective of the synthetic rock (synroc) and sodium zirconyl phosphate (NZP) strategy for high and intermediate level radioactive waste management is to provide a waste form, which has much greater resistance to leaching by ground water. In fact, the Synroc strategy for immobilization HLW is similar to the way the nature immobilizes radioactive elements on a scale vastly greater than will ever be contemplated by nuclear industry. Ceramic precursors are designed to contain about 20wt% or more of HLW calcine. It comprises an assemblage of four types of titanate phases: Hollandite, Zirconolite, Perovskite and Rutile (titanium oxide).

This present work is basically aimed at synthesis, solid state reactivity, crystallographic characterization and electrical measurements on titania and zirconia based ceramic materials, which have been identified as potential matrix for process development in technology of radioactive waste management. Though their applications in this area are well documented but their structure–property relationship is not yet investigated to the desired level, therefore, it is necessary to understand thoroughly the structural complexities of the interactions of the effluent cations vis-à-vis ceramic precursors. Powder diffraction data and GSAS software based calculations of cell parameters, crystal symmetry, isotropic and thermal parameters, interatomic distances and other structural factors provide a good data base for process development. The electrical properties of substituted ceramic phases have been investigated with a view to find their applications in electrical and electronic materials, insulators and sensor technology. Impedance spectroscopy is used as a prime technique for such investigations. It is in this context that the author has selected titanates, zirconates and phosphates for crystallochemical and electrical investigations.

1.7 Investigation of relevant literature

Metal Substituted sodium zirconium phosphate (NZP)

One of the most important materials that are being presently studied in materials science is sodium zirconium phosphate. The parent composition of $NaZr_2P_3O_{12}$ was first reported by Goodenough et al [1]. These materials have now attracted wide attention in academia and in industry for their potential as future candidates for devices requiring good thermal shock resistance, catalyst supports in the automobile industry, hosts for nuclear waste immobilization [2-3] and fast ionic conductor in sulfur battery [4-6]. NZP structure allows a huge variety of ion substitutions not only to the vacancies in the structure but also to the rigid framework [7]. The thermodynamic stability of NZP and its unique property of accommodating 40 to 45 elements in the periodic table in its lattice without altering the basic structure make it a potential matrix for nuclear waste immobilization [8-9]. $NaZr_2(PO_4)_3$ is a prototype for a broad family of compounds called "NZP". The structure of $NaZr_2(PO_4)_3$ consists of a three dimensional rigid framework which is stable but flexible. Furthermore, substitution to this framework allows hundreds of different compositions. NZP type compounds show several different properties; ionic conductivity, low and negative thermal expansion, catalytic activity, ferroelectricity, and the ability to immobilize radioactive nuclides. Some NZP structured materials that show negative thermal expansion find commercial applications as coatings [10] and physically and thermally stable, ultra low expansion, insulating ceramics. In addition, materials with high levels of ion exchangeability can find applications in the immobilization of undesired elements from nuclear and fuel reprocessing wastes [11-14] or as solid state electrolytes [15] or in high energy batteries etc. NZP-like structures are well known as excellent ionic conductors and recently tetravalent ion conduction was demonstrated in $MNb(PO_4)_3$ (M = Zr, Hf) [16-19].

The structure of $NaZr_2(PO_4)_3$ was reported as rhombohedral with the space group R-3c. The framework is composed of ZrO_6 octahedra and PO_4 tetrahedra that share corners through strong P-O-Zr bonds. In addition, counter ion Na^{+1} is located in the specific M1 site inside the framework. General formula for the NZP family is $M(I)_{0 \to 1} M(II)_{0 \to 3} M(III)_{0 \to 1} A_{2n}(XO_4)_{3n}$ in which $[A_{2n}(XO_4)_{3n}]^{m-}$ represents the rigid framework and M(I), M(II) and M(III) represent the three distinct empty sites in the framework. Depending on the composition of the material the symmetry may lower from R-3 c to R-3, C2/c, Cc, etc [20-21].

In the compounds with the highest symmetry R-3c, the M(I) site has 3 symmetry and the coordination number is 6. It lies on the c axis and generally this site is fully or partially occupied. The M(II) site has 2 symmetry and the coordination number is 14. It lies between the columns and is connected to the M(I) site allowing ionic conductivity. Furthermore, as the M(II) site is very big, in some cases two cations may fit into the site. The M(III) site is a distorted trigonal prism and has 32 symmetry. It can only accommodate very small ions and in most cases it is empty. M, A and X can be substituted with many different elements forming hundreds of different compositions [22]. **M** can be substituted with H^+, Li^+, Na^+, K^+, Rb^+, Cs^+, Cu^+, Ag^+, Ga^+, Tl^+, NH^+, Mg^{2+}, Ca^{2+}, Sr^{2+}, Ba^{2+}, Mn^{2+}, Co^{2+}, Ni^{2+}, Cu^{2+}, Zn^{2+}, Cd^{2+}, Hg^{2+}, Fe^{3+}, Ln^{3+}, Bi^{3+}, Zr^{4+}, Hf^{4+}, **A** can be substituted with Nb^{5+}, Ta^{5+}, V^{5+}, Sb^{5+}, Ti^{4+}, Zr^{4+}, Hf^{4+}, Ge^{4+}, Sn^{4+}, Mo^{4+}, U^{4+}, Np^{4+}, Pu^{4+}, Nb^{4+}, Sc^{3+}, Y^{3+}, Ln^{3+}, V^{3+}, Cr^{3+}, Fe^{3+}, Co^{3+}, Al^{3+}, Ga^{3+}, In^{3+}, Ti^{3+}, Mg^{2+}, Mn^{2+}, Cu^{2+}, Co^{2+}, Ni^{2+}, Zn^{2+}, Na^+, K^+, and **X** can be substituted with As^{5+}, Si^{4+}, Ge^{4+}, S^{6+}, Mo^{6+}, Al^{3+}. Different compositions give several different properties to the structure like ionic conductivity, low and negative thermal expansion and ability to immobilize radioactive nuclides, ferroelectricity and catalytic activity. Substituting M site ions can be done at relatively low temperatures, around 300°C, and the framework ions, A and X, can be done at higher temperatures, more than 800°C. As the oxidation state of A and X gets higher the framework becomes stronger and the number of different substitutions to the M site increases.

Inside of the framework, $[A_{2n}(XO_4)_{3n}]^{m-}$, M^{1+} counter ions fully occupy the M(I) vacancy, which is located on the c axis, between two ZrO_6 octahedra. In addition, as the M^{+1} ion gets bigger, the c axis expands and PO_4 tetrahedra distort. Distortion of PO_4 tetrahedra increases the O-P-O angle through c axis, which shortens the **a** axis.

The low thermal expansivity of NZP compounds was first observed by Boilot *et al.* [23]. Lenain et al. [24] have provided a structural model that explains this behavior, and the model has been verified by Hazen *et al.* [25] by means of high temperature X-ray difractometry. A large number of NZP compounds have been investigated in the search for low-thermal-expansion materials (26-28). Taylor [29] and Brewal and Agrawal [30] have reviewed the thermal expansion behavior of NZP compounds and compiled the available information. Limaye et al [31] have reported the synthesis and thermal expansion of a series of alkaline earth containing NZP compounds, namely $A_{1/2}Zr_2(PO_4)_3$ (A = Mg, Ca, Sr, or Ba). They found that $Ca_{1/2}Zr_2(PO_4)_3$ and $Sr_{1/2}Zr_2(PO_4)_3$ show reverse axial anisotropy as well as opposite bulk thermal expansivity.

Metal Substituted calcium zirconium phosphate (CZP)

One of the derivative materials that have been extensively studied is calcium zirconium phosphate [32-34] which is isostructural with Sodium zirconium phosphate. Calcium zirconium phosphate has high temperature thermal stability and excellent thermal shock resistance, which render the material to be a potential candidate for host of application such as radio active waste isolation, thermal shock barrier coating, auto mobile engine component and catalyst support. Calcium zirconium phosphate has very low average bulk thermal expansion and exhibit anisotropy in the thermal expansion, which lower the bulk thermal expansion of the polycrystalline material through micromechanical stress induced-microcracking during cooling from the processing temperature.

Calcium Zirconium phosphate has been widely studied with a view to their potential application as a catalyst, ion exchanger and ion conductor. The calcium diorthophosphate $CaZr(PO_4)_2$ was prepared in an earlier study by Bettinali et al [35], and the unit cell parameters were determined from x-ray diffraction data. However as pointed out by Kinoshita and Inoue [36], these seems to be some uncertainty in its chemical composition. Koichiro et al [37] describe crystal structure of calcium zirconium diorthophosphate. Calcium-strontium zirconium phosphates concern to orthophosphate group and are characterized by the wide development of isomorphic replacement of various groups of elements [38].

During last few years a new structural family of low thermal expansion materials namely sodium zirconium phosphate (NZP) and calcium zirconium phosphate (CZP) has been developed. The group is characterized by a flexible framework structure belonging to the rhombohedral system with possibility of isomorphic replacements of various groups of elements [39]. In recent years these solid solutions are receiving attention for their potential to be used as ionic conductors and host material for radioactive waste immobilization because of their structural flexibility with respect to isomorphic ionic replacements and high stability against leaching reactions [40]. Compounds with NZP skeleton are anisotropic, changing their dimension in apposite magnitudes when the counter ion of the skeleton is substituted or thermally affected. This fact is the basis of a series of materials with very low thermal expansion ($\alpha \sim 10^{-7}\,°C^{-1}$) [41-42]. It is this feature of the NZP and CZP skeleton which has generated interest in the study of mobility and resultant properties of these compounds. A model has been developed in order to put into simple parameters the atomic contributions and topological relations to the lattice dimensions. The structure of NZP compound in most cases belongs to the R-3c group whereas structure of CZP, Sr and Ba modified CZP fits into R-3. B. Srinivasulu et al [43] have reported the Preparation of a new family of NASICON type phosphates $Ca_{0.5}NbMP_3O_{12}$ (M = Fe, Al, Ga and In) and characterization of the iron systems by Mossbauer spectroscopy. The materials were prepared by a solid state reaction method. They have been characterized by X-ray diffraction and IR spectroscopy. Mossbauer spectra show that majority of trivalent iron ions in the parent material are reduced to Fe^{2+} ions although small but detectable amount Fe^{3+} is still present. Basavaraj Angadi [44] have studied the influence of 50 MeV Li^{3+} ion irradiation on the thermal expansion of the low thermal expansion ceramic $Ca_{1-x}Sr_xZr_4P_6O_{24}$ (x = 0:00, 0.25, 0.50, 0.75 and 1.00) belonging to the sodium zirconium phosphate (NZP) family of ceramics was in the temperature range 300–1100˚K. The XRD and scanning electron microscope (SEM) studies indicate that the ion irradiation causes amorphization, especially at the grain boundaries. The thermal expansion hysteresis reduces due to Sr substitution and is further reduced upon irradiation. Chong S. Yoon et al [45] reported Synthesis of low thermal expansion ceramics based on $CaZr_4(PO_4)_6$–Li_2O system. Ultra-low thermal expansion ceramics based upon $CaZr_4(PO_4)_6$–Li_2O was synthesized through solid state sintering. The microstructure of $CaZr_4(PO_4)_6$–Li_2O was sensitive to composition due to the high solubility of Li_2O in $CaZr_4(PO_4)_6$. O. Mentre and F. Abraham [46] have reported Structural study and conductivity properties of $Ca_{1-x}Na_{2x}Ti_4(PO_4)_6$ solid solution. The structure of $CaTi_4(PO_4)_6$ was refined in the R-3 space group applying the Rietveld method to the X-ray powder spectra. Author was point out that ionic conductivity is a little better for $CaTi_4(PO_4)_6$ than for $Na_2Ti_4(PO_4)_6$ despite of the higher valence of Ca^{2+} compared to Na^+. D.K. Agrawal and V.S. Stubican have reported synthesis and sintering of $Ca_{0.5}Zr_2P_3O_{12}$ – a low thermal expansion material [47]. The single phase compound $Ca_{0.5}Zr_2P_3O_{12}$ (CZP) was prepared by solid state reaction technique. This material shows a negative thermal expansion in the temperature region of 30°-500°C. The lattice thermal expansion of rhombohedral $CaZr_4(PO_4)_6$ has been determined to be anisotropic with a negative expansion coefficient along the directions in the plane (000).The high temperature neutron diffraction study of $CaZr_4(PO_4)_6$ has allowed the understanding of its anomalous thermal behavior. A.R. Kotelnikov et al [48] have reported synthesis and x-ray diffraction study of solid solutions $(Ca,Sr)_{0.5}Zr_2P_3O_{12}$. The materials were prepared by of sol-gel method. They have been characterized by X-ray diffraction and IR spectroscopy.

Metal Substituted perovskites

The mineral perovskite is named after a Russian mineralogist, Count Lev Aleksevich von Perovski, and was discovered and named by Gustav Rose in 1839 from samples found in the Ural Mountains. Since then considerable attention has been paid to the perovskite family of compositions.

Ceramic oxide materials such as perovskite ($CaTiO_3$), zirconilte $CaZrTi_2O_7$, hollandite ($Ba_{1.23}Al_{2.46}Ti_{5.54}O_{16}$), pyrochlore ($Ln_2Zr_2O_7$ and $Ln_2Ti_2O_7$, Ln = rare earth metals) and sphene ($CaTiSiO_5$) have gained tremendous interest because of their application as the geological medium for the immobilization of radioactive wastes [49-50]. The stacking of the metal-oxygen polyhedra in their structure results in the formation of cavities and vacant inter layers capable of accommodating a large number of radioactive cations. The most popular procedure employed by the nuclear power establishments during the past few decades has been to incorporate the nuclear wastes into borosilicate glasses. The serious disadvantage recognized recently is that the borosilicate glasses readily devitrified when subjected to action of water and steam at elevated temperatures and pressure [51]. Therefore an alternative method using ceramic materials in which the radionuclide are incorporated into solid solution in an assemblage of mineralogical phases is adopted by several workers [52-53]. Since the nuclear waste contains a variety of ions of various sizes and charge, it is difficult to incorporate all the radioactive ions in one phase. So a mixture of phases, i.e. a phase assemblage in which the phases are chemically compatible with one another, has been considered. A variety of phase assemblages have been proposed like super calcine [54-55] containing scheelite, cubic zirconia, spinel, apatite, corundum and pollucite phases. Synthetic rock (SYNROC), a titanate based ceramic containing hollandite, perovskite and zirconolite phases, has been investigated for nuclear waste immobilization. Fine crystals of sphene ($CaTiSiO_5$, calcium titanosilicate) in an alumino silicate glass matrix are also being studied as alternatives to glasses for the immobilization of nuclear wastes [56].

Substituted calcium titanate

The mineral perovskite $CaTiO_3$ is the parent compound of the perovskite-structured family. Calcium titanate exhibit high ionic conductivity, which leads to a wide variety of applications including solid oxide fuel cells, and their crystal structures have extensively been studied [57–66]. $CaTiO_3$ is commonly used as analog to the $(Mg,Fe)SiO_3$ [67], which is dominant phase in the lower mantle. $CaTiO_3$ has been recognized as one of the important constituents for immobilization of high-level radioactive wastes and is being widely used in electronic ceramic materials [68-69]. Calcium titanate structure is made up of corner sharing TiO_6 octahedra with Ca at the corner of the unit cell. Crystal structure and phase transition of the $CaTiO_3$ are of particular interest in materials science and in the earth sciences. Calcium titanate is a perovskite structure having orthorhombic symmetry [70-71]. Sasaki reported [72] that the structure was orthorhombic at room temperature and transforms to the cubic ideal perovskite at high temperature, with an intermediate tetragonal phase [73-75]. Also, phase transition from orthorhombic to tetragonal and cubic symmetry has been reported with increasing Fe content at room temperature in the system $CaFe_xTi_{1-x}O_{3-(x/2)}$ for $0.0 \leq x \leq 0.4$. Thus, for $x \leq 0.205$ the structures are orthorhombic and for $x \geq 0.251$ they are cubic, the intermediate compositions showing tetragonal symmetry [76]. Kim et al. [77] reported, that in $CaTiO_3$ the partial substitution of Ca^{2+} with La^{+3} leads to the formation of single phase $Ca_{1-x}La_{2x/3}TiO_3$ solid solution with structural changes from orthorhombic to tetragonal double layered perovskite as x increases. Extensive study [78-79] has been carried out on $CaTiO_3$ to improve its dielectric properties indicating that on replacing divalent calcium by trivalent Yttrium or other rare earth ions, the charge neutrality is maintained by creating an appropriate number of vacancies on calcium sites. On other hand, Lewis and Callow [80] have reported, that donor substitution on A-site will lead to the formation of titanium vacancies to maintain electrical charge neutrality, while the acceptor dopant such as K^+ and Fe^{3+} will generate oxygen vacancies [81]. The structural phase transitions in pure $CaTiO_3$ have been investigated by a lot of authors by various techniques: heat capacity measurements, Raman spectroscopy, high temperature X-ray diffractometry, differential thermal analysis [82-86]. Irrespective of the method used, two phase transitions were detected at very high temperatures. The first one occurs in the range 1373-1423°K and the second one appears at 1523 ± 10°K. These transformations correspond to the sequence of crystallographic changes orthorhombic (Pbnm) ↔ tetragonal (I4/mcm) ↔ cubic (Pm3m) [87].

Nowadays, advanced ceramics became the key of success for the development of integrated circuits in microelectronic industry (88-89). Calcium titanate could be of great usefulness in this field of applications in the future. Nevertheless, pure $CaTiO_3$ ceramics are difficult to densify and too fragile for many practical applications whereas compositionally modified ceramic bodies derived from calcium titanate could be of high interest in special applications such as capacitors and resonators. Calcium titanate is well known for the treatment of radioactive wastes since this perovskite forms a vast number of solid solutions with rare earth metals [90-93]. Calcium titanate is reported to be kinetically stable in extreme geo hydrothermal conditions (up to 900°C and 5000 bar) and are, therefore, more stable to leaching than borosilicate glasses [94-97]. It has also been pointed out that the perovskite phase has a characteristic capability of converting several rare earths into titanate solid solutions at temperatures between 700°C and 1000°C. In this context it is desirable to understand the structure property relationship of rare earth substituted perovskites of $M_xCa_{1-x}TiO_3$ compositions where M = rare earth [98-100]. On the other hand, $CaTiO_3$ based materials are used as catalysts for partial oxidation of light hydrocarbons [101-104].

References

[1] J. B. Goodenough, H.Y.P. Hong and J. A. Kafalas, Mat. Res. Bull. 11, **(1976)**: 203.

[2] R. Roy, E. R. Vance, and J. Alamo, Mat. Res. Bull. 17, **(1982)**: 585.

[3] Pratheep Kumar S., Buvaneswari, G., Raja Madhavan, R. Radiochemistry 53, **(2011)**: 421.

[4] H. Y. P. Hong, Mat. Res. Bull. 11, **(1976)**: 173.

[5] B. E. Taylor, A. D. English, T. Berzins, Mat. Res. Bull. 12, **(1977)**: 171.

[6] Yunhua Yang, Guangjian Dai, Shaozao Tan, Yingliang Liu, Qingshan Shi, Yousheng Ouyang. Journal of Rare Earths, 29[4], **(2011)**: 308.

[7] Alamo J., Solid State Ionics 63-65, **(1993)**: 547.

[8] A. H. Naik, N. V. Thakkar, S. R. Dharwadkar, K. D. Singh Mudher, V. Venugopal, Journal of Thermal Analysis and Calorimetry 78, **(2004)**: 707.

[9] W. Lutze and R.C. Ewing, Radioactive waste for the future. Elsevier Science Publication **(1988)**: 238.

[10] Lee. W. Y, Cooley, K. M., Berndt C. C., Joslin D. L, Stinton D. P, Journal of the American Ceramic Society 79(10), **(1996)**: 2759.

[11] I. W. Donald, B. L. Metcalfe, R. N. J. Taylor, J. Mater Sci. 32, **(1997)**, 5851

[12] Pet'kov, V. I. and Sukhanov, M. V. Czechoslovak, Journal of Physics 53, **(2003)**: A671.

[13] M. V. Sukhanov, V. I. Pet'kov and D. V. Firsov. Inorganic Materials, 47[6], **(2011)**: 674.

[14] A. H. Naik, S. B. Deb, A. B. Chalke, M. K. Saxena and K. L. Ramakumar, et al. Journal of Chemical Sciences, 122[1] **(2010)**: 71.

[15] Imanaka, N and Adachi, G.Y., Journal of Alloys and Compounds 250, **(1997)**: 492.

[16] Imanaka, N., Itaya, M.; Ueda, T. and Adachi, G., Solid State Ionics 154, **(2002)**: 319.

[17] Imanaka, N., Ueda, T., Okazaki, Y., Tamura, S.; Hiraiwa, M. and Adachi, G. Y., Chemistry Letters, 5, **(2000)**: 452.

[18] Hagman, L. O. and Kierkega.P, Acta Chemica Scandinavica 22(6), **(1968)**: 1822.

[19] N. Anantharamulu, K. Koteswara Rao, G. Rambabu, B. Vijaya Kumar and Velchuri Radha, et al. Journal of Materials Science, Vol. 46, Number 9, **(2011)**: 2821.

[20] Ingalls, R., Garcia, G. A. and Stern, E. A., Physical Review Letters 40, **(1978)**: 334.

[21] Pet'kov, V. I. and Orlova, A. I., Inorganic Materials 39, (10), **(2003)**: 1013.

[22] J. Alamo and R. Roy J., Amer. Ceram Soc. 67, **(1984)**: C78.

[23] J. P. Boilot, J. P. Salanie, G. Desplanches, D. Lepotier, Mater. Res. Bull. 14, **(1979)**: 1469.

[24] Lenain G E, McKinstry H A, Alamo J and Agrawal D K., J. Mater. Sci. 22 **(1987)**: 17.

[25] R.M. Hazem, L.W. Finger, D.K. Agrawal, H.A. Mckinstry and A.J Perrotta., J. Mater. Res; 2, **(1987)**: 329.

[26] M. V. Sukhanov, V. I. Pet'kov, D. V. Firsov, V. S. Kurazhkovskaya and E. Yu. Borovikova. Russian Journal of Inorganic Chemistry, 56[9], **(2011)**: 1351.

[27] V. I. Pet'kov, D. V. Firsov, A. V. Markin, M. V. Sukhanov and N. N. Smirnova. Inorganic Materials, 47[2], **(2011)**: 178.

[28] U. Ahmadu, A. O. Musa, S. A. Jonah and N. Rabiu. Journal of Thermal Analysis and Calorimetry, 101[1], **(2010)**: 175.

[29] D. Taylor Brit. Ceram. trans. J. 90, **(1991)**: 64.

[30] E. Brewal and D.K. Agarwal Brit. Ceram. trans. 94, **(1995)**: 27.

[31] S. Y. Limaye, D.K. Agrawal and H. A. Mckinstry., J. Amer. Ceram. Soc. 70, **(1987)**: C-232.

[32] E. Breval, H. Mikinstry, D K Agarwal, J.Am.Ceram.Soc.81 [4], **(1998)**: 926.

[33] S.Y.Limaye, D K Agarwal, R. Roy. J.Mater.Sci.26, **(1991)**: 93.

[34] D.A. Hirschfeld, T. K. Li., W. M. Russ, K. H. Lee., American ceramic society, V52, **(1995)**: 19.

[35] Bettinali, C., La Ginestra, A., and Valigi, M. Lincei, Classe sci, Fis, Mat. Nat. 33, **(1962)**: 472.

[36] Kinshita, M., and Inoue, Mnippon Kagaku Kaishi 8, **(1980)**: 1219.

[37] Koichiro Fukuda and Kazuko Fukutani, Powder Diffraction 18[4], **(2003)**: 296.

[38] Volkov Ju.F., Orlova A.I. Formulue types. Radiochemistry 38[1], **(1996)**: 15.

[39] Chourasia Rashmi and O.P. Shrivastava, Solid State Sciences 13 **(2011)**: 444.

[40] T. Oota, and I. Yamai, J. Am. Ceram. Soc. 69, **(1986)**: 1.

[41] R. Roy, D.K. Agrawal, J. Alamo, R.A. Roy, Mater. Res. Bull. 19, **(1984)**: 471.

[42] G.E. Lenain, H.A. McKinstry, S.Y. Iimaye, A. Woodward, Mater. Res. Bull. 19, **(1984)**: 1451.

[43] B. Srinivasulu, M. Vithal J. Mater. Sci. Latter 18, **(1999)**: 1771.

[44] Basavaraj Angadi, V.M. Jali, M.T. Lagare, N.S. Kini, A.M. Umarji, Ravi Kumar, S.K. Arora, D. Kanjilal, Nuclear Instruments and Methods in Physics Research B 187, **(2002)**: 87.

[45] Chong S. Yoon, Jae H. Kim, Chang K. Kim, K.S. Hong, Materials Science and Engineering B79, **(2001)**: 6.

[46] O. Mentre, F. Abraham., Solid State Ionics 72, **(1994)**: 293.

[47] D.K. Agrawal and V.S. Stubican, Mat. Res, Bull.20, **(1985)**: 99.

[48] A. R. Kotelnikov, A. M. Kovalsky, I. G. Trubach, A. I. Orlova, V. I. Petkov, Herald DGGGMS RAS, 5 (15), (2000) v.2.

[49] Clarke D Rann. Rev. Mater. Sci. 13, **(1983)**: 191.

[50] Hayward P. J. and Cecchetto E. V., in Scientific basis for nuclear waste management (ed) Z Lutze (New York: Elsevier) 5, **(1984)**: 91.

[51] MacCarthy G. J., White W. B., Rustum Roy, Scheetz B. E., Komarneni S., Smith D. K. and Roy D. M., Nature (London) 273, **(1978)**: 216.

[52] MacCarthy G J and Davidson M. T., Am. Ceram. Soc. Bull. 54, **(1975)**: 782.

[53] Ringwood A. E., Kessoa S. E., Ware N.G., Hibberson W. and Major A., Nature (London) 278, **(1979)**: 219.

[54] MacCarthy G. J. Trans. Am. Nucl. Soc. 23 **(1976)**: 168.

[55] MacCarthy G. Jin, Scientific basis for nuclear waste management (ed) (New York: Plenum) 1 **(1979)**: 329.

[56] Hayward P. J., Vance E. R., Cann C. D. and Mitchell S. L., in Advance in ceramics (ed) G. G. Wicks and W. A. Ross (Ohio: American Ceramic Society, Columbus) 8, **(1984)**: 291.

[57] T. Ishihara, T. Kudo, H. Matsuda, Y. Takita, J. Electrochem. Soc. 142, **(1995)**: 1519.

[58] K. Huang, R. Tichy, J.B. Goodenough, C. Milliken, J. Am. Ceram. Soc. 81, **(1998)**: 2581.

[59] T. Ishihara, T. Akbay, H. Furutani, Y. Takita,, Solid State Ionics 113–115, **(1998)**: 585.

[60] M. Yashima, K. Nomura, H. Kageyama, Y. Miyazaki, N. Chitose, K. Adachi, Chem.Phys. Lett. 380, **(2003)**: 391.

[61] H. Yoshioka, J. Am. Ceram. Soc. 85, **(2002)**: 1339.

[62] R. Ali, M. Yashima, F. Izumi, Chem. Mater. 19, **(2007)**: 3260.

[63] Y. Teraoka, H.M. Zhang, K. Okamoto, H. Yamazoe, Mater. Res. Bull. 23, **(1998)**: 51.

[64] M. Yashima, T. Tsuji, J. Appl. Crystallogr. 40, **(2007)**: 1166.

[65] M. Yashima, T. Kamioika, Solid State Ionics 178, **(2008)**: 1939.

[66] M. Yashima, M. Enoki, T. Wakita, R. Ali, Y. Matsushita, F. Izumi, T. Ishihara, J. Am.Chem. Soc. 130, **(2008)**: 2762.

[67] B. Gautason, K. Muehlenbachs, Science 260, **(1993)**: 518.

[68] Sarabjit Singh, Chandra Prakash, K.K. Raina Journal of Alloys and Compounds, Vol. 492, Issues 1-2, **(2010)**: 717.

[69] A.E. Ringwood, S.E. Kesson, K.D. Reeve, D.M. Levins, E.J. Ramm, in: W. Lutze, R.C.Ewing (Eds.), Radioactive Waste Forms for the Future, Elsevier, Amsterdam, **(1988)**: 233.

[70] H. F. Kay and B. C. Bailey, Acta. Crystallogr. 10, **(1957)**: 219.

[71] Rashmi Chourasia and O.P. Shrivastava. Bulletin of Materials Science, Vol. 34[1], **(2011)**: 89.

[72] S. Sasaki, Acta Crytallogr., C-43, **(1987)**: 1668.

[73] S. A. T. Redfern, J. Phys.: Condens. Matter, 8, **(1996)**: 8267.

[74] F. Guyot, P. Richet, Ph. Courtial and Ph. Gillet, Phys Chem Miner., 20, **(1993)**: 141.

[75] B. J. Kennedy, C. J. Howarad and B. C. Chakoumakos, J. Phys. Condens. Matter, 11, **(1999)**: 1479.

[76] A. I. Becerro, F. Seifert, R. J. Angel, S. Rios and McCammon, J. Phys. Condens. Matter, 12, **(2000)**: 3661.

[77] I.S. Kim, W. H. Jung, Y. Inaguma, I. Nkanura and M. Iath, Mater. Bull., 30(3), **(1995)**: 307.

[78] S. Neirman and I Burn, J. Mater. Sci., 19, **(1984)**: 737.

[79] U. Blalachandran and N. G. Eror, J. Mater. Sci.23, **(1988)**: 2676.

[80] G. V. Lewis and C. R. A. Callow, J. Phys. Chem. Solids, 47, **(1986)**: 89.

[81] B. Jaffe, J.W.R. Cook and H. Jaffe, "Piezoelectric Ceramics" Academic Press, New York, Chap. 5, **(1971)**.

[82] Gillet P., Guyot F., Price G. D., Tournerie P., Le Cleach A.: Phys.Chem.Miner. *20*, **(1993)**: 159.

[83] Matsui T., Shige matsu H., Arita H., Hanajiri Y., Nakamitsu N., Nagasaki T.: J. Nucl. Mater. 247, **(1997)**: 72.

[84] Sining Yun, Xiaoli Wang, Juanfei Li, Jing Shi, Delong Xu. Materials Chemistry and Physics,116[2-3], **(2009)**: 339.

[85] A. A. Murashkina, A. N. Demina, E. A. Filonova, A. K. Demin and I. S. Korobitsin Inorganic Materials, 44[3], **(2008)**: 296.

[86] S. Jesurani, S. Kanagesan, R. Velmurugan and T. Kalaivani. Journal of Materials Science: Materials in Electronics, 22, **(2011)**: 1-7.

[87] Redfern S.A.T., J. Phys.Condens.Mater. *8*, **(1996)**: 8267.

[88] Peng Hu, Huan Jiao, Chun-Hai Wang, Xiao-Ming Wang, Shi Ye, Xi-Ping Jing, Fei Zhao, Zhen-Xin Yue. Materials Science and Engineering B, 176[5], **(2011)**: 401.

[89] Xueqin Jin, Dazhi Sun, Mingjun Zhang, Yudan Zhu and Juanjuan Qian Journal of Electro ceramics, 22[1-3], **(2009)**: 285.

[90] McCarty G.J., Nucl. Technol. *32*, **(1977)**: 92.

[91] Roy R., Radioact. Waste Disposal *1*, **(1962)**: 36.

[92] Vance E.R., Day R.A., Zhang Z., Begg B.D., Ball C.J.,Blackford G.: J. Solid State Chem. *124*, **(1996)**: 77.

[93] Hanajiri Y., yokoi H., Matsui T., Arita Y. Nagasaki T., Shigematsu H.: J.Nucl.Mater. *247*, **(1997)**: 285.

[94] Cheary R. W. Acta Crystallogr. B42, **(1986)**: 225.

[95] Kesson S. E. and White T. J., Proc. R. Soc. London A405, **(1986)**: 73.

[96] Cheary R. W. and Squadrito R. M., Acta Crystallogr. B45, **(1989)**: 205.

[97] Gatehouse B M, Grey R J, Hill and Rossel *Acta Crystallogr*.B37, **(1989)**: 20.

[98] Ringwood A. E. "Safe disposal of high level nuclear reactor waste; A new strategy" (Canberra: Austral Nat Univ. Press) **(1978)**: 64.

[99] Myhra S, Giorgi R, Delogu P and Riviere J .C. J. Mater. Sci. 23, **(1986)**: 1514.

[100] Andersen A.G., Hayakawa T., Tsunoda T., Orita H., Shimizu M., Takehira K.: Catalysis Lett. *18*, **(1993)**: 37.

[101] A. V. Shyakhtina, A. V. Levchenko, J. C. C. Abrantes, V. Yu. Bychkov, V. N. Korchak, V. A. Rassulov, L. L. Larina, O. K. Karyagina, L. G. Shcherbakova, Mater. Res. Bull. 42, **(2007)**: 742.

[102] Vishal Singh, K.K. Bamzai, Shivani Suri, and Nidhi Ceramics International 37 **(2011)**: 2655.

[103] Satyendra Singh, S.B. Krupanidhi, Materials Chemistry and Physics 131 **(2011)**: 443.

[104] S. Jesurani, S. Kanagesan, R. Velmurugan, C. Thirupathi, M. Sivakumar, T. Kalaivani, Materials Letters 65 **(2011)**: 3305.

Chapter 2

Crystal Structure Refinement and instrumentation

2.1. Powder Diffraction method

A major emphasis of materials science is in understanding the elemental compositions and corresponding atomic structures present in materials of interest. This knowledge confirms a material's purity and suitability for use, and allows explanation for its properties and performance. Just as chemical elements form a plethora of compounds, so a compound may pack in different arrays to form a variety of distinct crystal structures (known as polymorphs or phases). Elemental composition and physical characteristics such as color and hardness might differentiate phases when encountered in pure form. When in mixtures or reacted with other materials, identification of phases based on physical characteristics or elemental composition can quickly become impossible.

Powder diffraction has been the staple analytical tool for chemists and materials scientists for more than 50 years. Powder diffraction is a tool to identify and characterize materials by analyzing the radiation scattering produced when the materials are illuminated with X-rays or neutrons. The patterns formed by the scattered radiation provide an abundance of information from simple fingerprinting to complex structural analysis. X-ray powder diffraction is a powerful non-destructive testing method for determining a range of physical and chemical characteristics of materials. It is widely used in all fields of science and technology [1]. The applications include phase analysis, i.e. the type and quantities of phases present in the sample, the crystallographic unit cell and crystal structure, crystallographic texture, crystalline size, macro-stress and micro strain and also electron radial distribution functions.

The usefulness of powder diffraction ranges throughout all areas where materials occur in the crystalline solid state.

Uses for powder diffraction are found within the following fields and beyond:

- Natural Sciences
- Materials Science
- Pharmaceuticals
- Geology and Petrochemicals
- Engineering
- Metallurgy
- Forensics
- Conservation and Archaeology

The term "powder", as used in powder diffraction, does not strictly correspond to the usual sense in the word in common language. In powder diffraction the specimen can be a "solid substance divided into very small particles" But it can also be a solid block for example of metal, ceramic, polymer, glass or even a thin film or a liquid. The reason for this is that the important parameters for defining the concept of a powder for a diffraction experiment are the number and size of the individual crystallites that form the specimen, and not their degree of accretion [2]. An "ideal" powder for a diffraction experiment consists of a large number of small, randomly oriented crystallites (coherently diffracting crystalline domains). If the number is sufficiently large, there are always enough crystallites in any diffracting orientation to give reproducible diffraction patterns.

2.2. Determination of crystal structure

Atomic structure is the most important piece of information about crystalline solids: just from the knowledge of topology of the structure, a precise structural model and many physical properties of crystals can be calculated with state of-the-art quantum-mechanical methods.

"The ability to determine crystal structures directly from powder diffraction data promises to open up many new avenues of structural science. Many important materials cannot be prepared as single crystals of appropriate size and quality for conventional single crystal diffraction studies, nor indeed for the emerging synchrotron-based microcrystal diffraction techniques. In such cases, structure determination from powder diffraction data may represent the only viable approach for obtaining an understanding of the structural properties of the material of interest [3]". However, it is important to recognize that structure determination from powder diffraction data is far from routine and significant challenges must be overcome in developing and applying methods for this purpose. For this reason, several research groups have devoted considerable effort in recent years to the development of new and improved techniques in this field. More detailed reviews covering all aspects of structure determination from powder diffraction data may be found in references [4-9]. Crystal structure determination from diffraction data (either single crystal or powder) can be divided into the following stages: (i) unit cell determination and space group assignment, (ii) structure solution, and (iii) structure refinement. The aim of structure solution is to derive an initial approximation to the structure from direct consideration of the experimental diffraction data, but starting from no knowledge of the actual arrangement of atoms or molecules within the unit cell. If the structure solution is a sufficiently good approximation to the true structure, a good quality structure may then be obtained by structure refinement. For powder diffraction data, structure refinement is now carried out fairly routinely using the Rietveld profile refinement technique [10-11], and unit cell determination is carried out using standard indexing procedures (see, for example, references [12-17]. The techniques currently available for structure solution from powder diffraction data can be subdivided into two categories-"traditional" and "direct-space" approaches. As discussed below,

2.2.1.. Conventional approaches

In the traditional approach, the intensities I(hkl) of individual reflections are extracted directly from the powder diffraction pattern, and the structure is then solved using these I(hkl) data in the types of structure solution calculation that are used for single crystal diffraction data (e.g. direct methods or Patterson methods). However, as there is usually extensive peak overlap in the powder diffraction pattern, extracting reliable values of the intensities I(hkl) of the individual diffraction maxima can be problematic, and may lead to difficulties in subsequent attempts to solve the structure using these "single-crystal-like" approaches. To overcome this problem either requires improved techniques for extracting and utilizing peak intensities, or requires the use of new structure solution strategies that allow the experimental powder diffraction profile to be used directly in its "raw" digitized form, without the requirement to extract the intensities and (hkl) of individual diffraction maxima.

2.2.2. Straight space approaches

In the direct-space approach, trial structures are generated in direct space, independently of the experimental powder diffraction data, with the suitability of each trial structure assessed by direct comparison between the powder diffraction pattern calculated for the trial structure and the experimental powder diffraction pattern. This comparison is quantified using an appropriate R factor. Most direct-space approaches reported to date have used the weighted powder profile R factor R_{wp} (the R-factor normally employed in Rietveld refinement), although we note that some implementations of direct-space approaches have instead used R-factors based on extracted peak intensities.

The basis of the direct-space strategy for structure solution is to find the trial crystal structure corresponding to lowest R-factor, and is equivalent to exploring a hypersurface R (Γ) to find the global minimum, where Γ (capital gama) represents the set of variables that define the structure. In principle, any technique for global optimization may be used to find the lowest point on the R (Γ) hyper surface, and much success has been achieved in using Monte Carlo [18-23], Simulated Annealing [24-30] and Genetic Algorithm [31-36] methods in this field. In addition, grid search methods have also been employed [37-40]. This article focuses on fundamental and applied aspects of our implementations of Monte Carlo (MC) and Genetic Algorithm (GA) techniques within direct space structure solution from powder diffraction data, with particular emphasis on the application of these techniques to elucidate structural properties of molecular materials.

2.3. Refinement of crystal structure

Refinement is a general term that refers to almost all the operations needed to develop a trial model into one that best represents the observed data. Just as there is not a well defined mathematical technique for extracting valid phases from the observed intensities, so also there is no single well defined path from the trial model to the completed structure – if there were, it would have been programmed long ago. There are, however, some well trodden trails to guide a structure analyst, together with a growing number of validation tools. Refinement is a step-wise procedure, with increasingly subtle features being introduced in order to develop the model. Physical and chemical validation is a key feature of every stage of a refinement. In very difficult cases it is rare that mathematics alone will lead to an acceptable structure. In these cases knowledge of the chemistry or physical properties of the material may help to resolve uncertainties. Difficulties in programming 'scientific experience' have prevented the development of fully automatic structure analysis systems, though modern programs are beginning to bring 'do-it-yourself'' structure analyses into the hands of non specialists.

2.3.1. Rietveld Refinement Method

Rietveld refinement is a technique devised by Hugo Rietveld for use in the characterization of crystalline materials. The neutron and x-ray diffraction of powder samples results in a pattern characterized by reflections (peaks in intensity) at certain positions. The height, width and position of these reflections can be used to determine many aspects of the materials structure. The Rietveld method uses a least squares approach to refine a theoretical line profile until it matches the measured profile. The introduction of this technique was a significant step forward in the diffraction analysis of powder samples as, unlike other techniques at that time; it was able to deal reliably with strongly overlapping reflections.

"The Rietveld method is a complex minimization procedure. It is not an active tool for *ab initio* crystal structure analysis. It can only slightly modify a preconceived model built on external previous knowledge. The starting parameters for such a model must be reasonable close to the final values. More over, the sequence into which the different parameters are being refined needs to be carefully studied. The Rietveld method is a structure refinement procedure. It uses step intensity data $y(i)$, whereby each data point is treated as an observation".

The idea behind the Rietveld method is to consider the entire powder diffraction pattern using a variety of refinable parameters. That way the intrinsic problem of any powder diffraction pattern with its systematic and accidental peak overlaps is overcome. It is the intention to extract as much information as possible from a powder pattern. At the beginning this was restricted to atomic positions from neutron diffraction patterns. The first full publication has the title "*A Profile Refinement Method for Nuclear and Magnetic Structures*" [10]. Rietveld describes a structure refinement method (and program) which is not based on integrated intensities, as is done in single crystal structure refinements. His program employs directly the individual intensities $y(i)$ at each 2θ value obtained from step-scanning measurements of powder diffraction patterns. This was a novel and unusual approach. Rietveld considered such an approach and such a program since investigations based on powder methods had gained new importance, especially in neutron diffraction, owing to the general lack of large specimens for single crystal methods.

Diffraction of a polycrystalline sample, with neutrons or X-rays, reduces the three dimensional reciprocal lattice to a one-dimensional diagram. As a consequence such patterns suffer from overlapping peaks, sometimes accidental due to a lack of resolution, sometimes intrinsic in patterns for samples with cubic or trigonal symmetry.

Especially at higher diffraction angles, in low symmetry structures or when large unit cells are involved this is a serious problem. It is inevitable that certain information is lost. By using the step-scanned profile intensities instead of integrated intensities in the refinement procedure this difficulty can be overcome to a great extent allowing at the same time the extraction of a maximum amount of information. Rietveld's approach is based on the complete diffraction pattern, including background. The method is called originally "full pattern refinement". Since the IUCr satellite meeting on powder diffraction in Krakow, Poland, in 1978 terms like "Rietveld refinement", "Rietveld method" or "Rietveld analysis" were suggested. Rietveld method is generally and widely used today and extended to more and more fields of analysis based on diffraction, like phase analysis or texture analysis. This approach is fundamentally different from the Two Stage method based on POWLS, which was written with the same scientific background in the early sixties before the Rietveld program became known. POWLS requires intensities, as peak maxima or as integrated intensities. Later in the eighties the Two Stage method was extended by adding as a first step profile fitting resulting in a complete separation of the diffraction peaks. This step yields reliable integrated intensities comparable to single crystal data, and this makes this method more general and closer to single crystal analyses. This method consequently is called today the "Two Stage method".

There is much more information hidden in a powder pattern than just atomic positions, site occupancies and Debye-Waller factors. To name a few: lattice parameters and space group can be deduced and refined from the peak positions of the reflections; the amorphous fraction in the specimen or local order/disorder can be deduced from the background; particle size, strain/stress and domain size in the sample from analyzing the broadening of the peaks, *FWHM*, and in the recent developments qualitative and quantitative phase analysis. Special version has been written and numerous versions of the program are available today.

As input information the program requires:
- Initial values of all variable parameters (listed above)
- Step-scan data in equal increments 2θ
- 2θ limits, starting and ending values of 2θ, and regions which shall be excluded in the data
- Wavelength data

The parameters that can be adjusted in the least squares refinement, in principle simultaneously include:
- Lattice parameters ($a, b, c, \alpha, \beta, \gamma$)
- Atomic positions (x, y, z)
- Atomic site occupancies
- Atomic thermal vibrational parameters, isotropic or anisotropic
- Profile including U, V, W from the Cagliotti formula and asymmetry
- Preferred orientation
- Background function
- 2θ -zero correction
- Overall scale factor

Criteria for a successful refinement are the following:
- A difference plot y_i(obs) – y_i(calc)
- No maximum deviation at any point in the difference plot
- low Residual factors
- Structural parameters and their standard deviations (if possible in comparison to similar single crystal results)

R-factors in general are of limited values in judging the results obtained. Powder diffraction data are not directly comparable to single crystal data, but the R-factors used are close to the single crystal structure R-value. Also the standard deviations σ cannot be taken at face value. Of the several R-factors used to check on the quality of the refinement. Rwp is statistically the most meaningful indicator since the numerator is the residual that is minimized in the least squares procedure. Any R-factor presented is meaningful only if the background has been subtracted. This deserves a word of caution.

All R-factors are greatly affected, if the crystallites are not ideally "imperfect", e.g. if they are affected by crystallite size and/or micro strain problems. Crystallite-size shows up in line broadening and produces intrinsic contributions to the Lorentz profile. Micro strain on the other side produce Gaussian shaped profiles. Such contributions will show up in general in the difference plots. Very general the R-factors as quantities for fit and accuracy must be taken with care.

2.3.2. Rf factors

"The quality of the model can be judged with the help of various residual factors or 'R-factors'. These factors should converge to a minimum during the refinement and are to be quoted when a structure is published. The three most commonly used residual factors are:

Quality of each trial structure is assessed by the R factors. Some often used numerical criteria of fit are as follows:

(1) R-pattern (Rp)

$$R_p = \frac{\sum y_i(obs) - y_i(cal)}{\sum y_i(obs)}$$

(2) R-expected (Rexp)

$$R_e = \left[(N-P)\Big/\sum w_i y_{oi}^2\right]^{1/2}$$

(3) R-weighted pattern (Rwp)

$$R_{wp} = \left\{\frac{\sum w_i(y_i(obs) - y_i(cal))^2}{\sum w_i(y_i(obs))^2}\right\}^2$$

Where $y_{i(o)}$ and $y_{i(c)}$ are observed and calculated intensities at profile point i, respectively, N and P shows the numbers of data and of refined parameters and w_i is a weight for each step i respectively. R_{wp} considers the whole digitized intensity profile point-by-point, rather than the integrated intensities of individual diffraction maxima.

Thus, R_{wp} implicitly takes care of peak overlap and uses the digitized powder diffraction data directly "as measured".

Finally, there is the goodness of fit: GoF or simply S.

$$S = R_{wp}/R_{exp}$$

Theoretically, for a properly adjusted weighting scheme, the value for S should be close to 1. However, manipulating or rescaling the weights w can artificially improve this value. A goodness of fit of $S < 1$ suggests the model is better than the data. Obviously this is suspicious and usually a sign that there are some problems with the data and/or the refinement. Frequently, failure to perform a proper absorption correction leads to underestimated GoF values, but refinement in the wrong space group can also have this effect".

The strategy for structure solution in the direct-space approach is to find the trial structure that has the lowest possible R factor. However, effort to search trial solutions in minimum energy state has rarely been made in this field. In the present study, structure determination of the ceramic materials from powder X-ray diffraction pattern was carried out to generate trial structures in direct space. Cell constants of ceramic phases were determined from powder X-ray diffraction pattern by using CRISFIRE software.

2.4. Application of Different software for solving crystal structure

2.4.1. Indexing programs

In order to solve a crystal structure from powder diffraction data as the first step the pattern must be indexed. About 50–90% of all structure determination attempts fail because of failing indexing. Scientists devoted their time very early to this problem and a number of programs are available to help in indexing a diffraction pattern. See for example DICVOL91 by Louer [41]; TREOR by Werner [42--43]. Those algorithms have served the scientific community for decades; and they are adequate for most indexing problems, especially for volumes less than 1000 A^3. Since the 1990s apparently no significant progress has been made in developing new and especially more straight forward and more reliable indexing algorithms. This situation has changed in the last years, when a number of programs have been published, for example by Coelho (2003). Nearly all these mature indexing programs have their roots in the 1970s.

In practice indexing programs will never provide a single and unambiguous solution. Indexing is not a black-box approach; solid crystallographic background is a requirement. Use of several programs can maximize the possibility to find, and also to identify a correct solution. Taking advantage of different algorithms gives some idea of the range of different solutions and helps if identical solutions or derivative cells can be identified. Common requirements for successful indexing are sample purity and line-position data of the highest possible quality. The most popular indexing programs today probably are ITO, TREOR, and DICVOL. ITO is a deductive search program by zone-indexing in index space; All of them are included in a program; CRYSFIRE, which is developed by Robin Shirley [44]. It performs best when given 30 to 40 accurately measured powder lines. It is optimized for low symmetry systems (orthorhombic downwards); high-symmetry lattices may get reported wrongly in an orthorhombic setting. TREOR is a semi-exhaustive, heuristic search method in index space; it requires about 25 accurately measured powder lines. It is effective for searches down to triclinic symmetry.

DICVOL again is an exhaustive search program in parameter space by successive dichotomy; it requires around 20 accurately measured powder lines and is well suited for high symmetry down to mono clinic, impurity lines however are not tolerated. In all cases the success depends on accurate peak positions.

Recently more advances in powder indexing have been provided by Siemens/Bruker in a system called the SVD-index method and LP-Search. SVD-indexing (SVD = Single Value Decomposition) operates on d-values extracted from reasonable quality powder diffraction data. It can operate as an iterative process with hkl's assigned, or by a Monte Carlo approach searching the parameter space. It is not an exhaustive method. The programs ITO, TREOR and DICVOL have been incorporated into this package. Another approach, in the same package, is LP-Search [45], a Monte Carlo based whole powder pattern decomposition program. It is independent of d-spacing extraction and is therefore suited for indexing of poor quality powder data. No d-values are required as input data.

After indexing the pattern the actual analysis for determining crystal structures form powder diffraction data has to begin with profile fitting and profile refinement techniques. For this the peak shape must be known. The better the reflections can be described the better are the results afterwards. One serious problem is that the profiles in general change with diffraction angle. A great number of crystal structures, even complex crystal structures have been studied in recent years. Even if "Rietveld refinement" appears in the title, in reality a structural model has been deduced before by "conventional" methods, and this model was then refined by a "Rietveld" program.

Determination of a crystal structure from powder diffraction data follows the following steps.

- Collection of a highly resolved powder diffraction pattern.
- Indexing the powder pattern, determination of the unit cell and space group.
- Determination of intensities, resulting in a list of *hkl* and *I* (Miller indices and intensities). For solving unknown structures the intensities may be of limited accuracy.
- Structure solution using for example Patterson methods, direct methods or any other general approach.
- Structure refinement by "Rietveld".

2.4.2. General Structure analysis system software

The GSAS package has been developed over a period of decades by Allen C. Larson and Robert B. Von Dreele [46]. It is widely used and is arguably one of the most comprehensive packages of crystallographic software ever developed, as it can be used to fit crystallographic and magnetic structural models to X-ray and neutron single-crystal and powder diffraction data. Up to 99 sets of data may be used in combination to determine a single crystallographic model. In powder diffraction, it is common for a sample to contain more than one crystallographic phase. GSAS allows a crystallographic model to be composed of as many as nine crystallographic phases.

"In addition to structure determination, GSAS is also used for lattice-constant determination (even for materials with unknown structures), simulation of powder diffraction data, and for texture analysis. Recently, it was enhanced to enable the fitting of protein structures to powder diffraction data. The GSAS package contains approximately fifty programs and each of these programs is designed for specific types of task or type of crystallographic calculations. The majority of the programs require minimal or no user input. The only program in the GSAS package that requires extensive user interaction is named EXPEDT, an acronym for experiment editor. The GSAS package is quite sophisticated because a lot of options are available. The alternative method to execute commands and control options in GSAS is a program,

EXPGUI. This is a graphical user interface (GUI) editor for GSAS experiment (EXP) files and shell that allows all the other GSAS programs to be executed with a GUI. Graphical user interfaces (GUIs) simplify use of computers by presenting information in a manner that allows rapid assimilation and manipulation. The use of visual constructs that mimic physical objects such as `switches' and `buttons' can speed learning, by providing an intuitive method to provide input to the computer. A GUI is not always an improvement. As demonstrated by some widely used commercial programs, a poor GUI implementation can obscure functionality. If the GUI is organized in a counterintuitive manner, or if the menu contents are arranged haphazardly, or if commonly performed operations require several unexpected steps to be performed, then a user must typically invest a significant amount of time in learning how to use the program before the program can be used effectively. The EXPGUI program offers an alternate, GUI-based mechanism for reviewing and editing the GSAS experiment file. The EXPGUI program implements only a small fraction of the full complement of features available in EXPEDT, but most of the commonly used features for powder diffraction are present in EXPGUI. Since EXPGUI and EXPEDT can be used interchangeably, use of EXPGUI does not prevent access to any capabilities within GSAS. The EXPGUI program can be used in two ways. It can be run from one of the several platform-specific `shell programs' that allow access to the various GSAS programs. More significantly, EXPGUI can function both as an experiment file editor and as a GSAS shell program, providing GUI access to initiate other GSAS programs [47]. In the present study, EXPGUI was used to control three main programs: POWPREF, GENLES and LIVEPLOT in GSAS. (1) POWPREF prepares powder diffraction data for subsequent least squares analysis. (2) GENLES is a least squares refinement program. It constructs a single full least squares matrix and vector using multiple data sets. A mixture of powder diffraction and single crystal data for a given structural problem can thus be processed simultaneously. (3) LIVEPLOT draws the experimental, the simulated patterns and the difference between two patterns together. It performs the following operations of structural refinement".

POWPLOT	Powder pattern plotting
POEPREF	Powder data preparation
GENLES	General least square
DISGAL	Bond distance and angle calculation
FOURIER	Fourier map calculation
VRSTPLOT	Preparation of virtual reality crystal file
ORTEP	Crystal structure plotting

It is capable of handling all of these types of data simultaneously for a given compound. In addition, it can handle powder diffraction data from a mixture of phases refining structural parameters for each phase. Quantitative Rietveld details of phase analysis are available in the standard text books of crystallography [48]. The overall functional sequence of GSAS programming can be summarized by the following flow sheet.

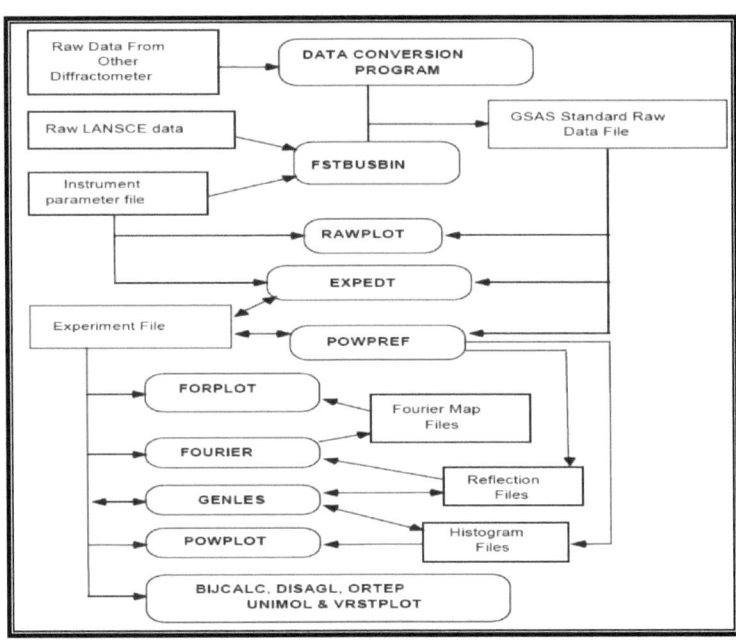

2.5. Scanning electron microscopy

Scanning electron microscopy (SEM) is a widely used technique employed to image the surface of samples. It provides out standing image resolution, unique image contrast and a large depth of field. Additionally, SEM often requires little sample preparation and can be coupled with emerge dispersive x-ray spectrometer (EDAX) to provide elemental identification. These attributes make SEM an outstanding choice for imaging application that require the resolution greater then that provided by the optical microscopy or when there is a need to identify elemental continuants of micrometer–sized features. Sample preparation methods for SEM are highly developed for the imaging of sub-surface features. Method include chemical etching, mechanical cross sectioning, planer polishing to remove layers parallel to surface and ion beam technique. When the electron beam hits the sample, the interaction of the beam electrons with the sample atoms generates a variety of signals (Fig. 2.1).

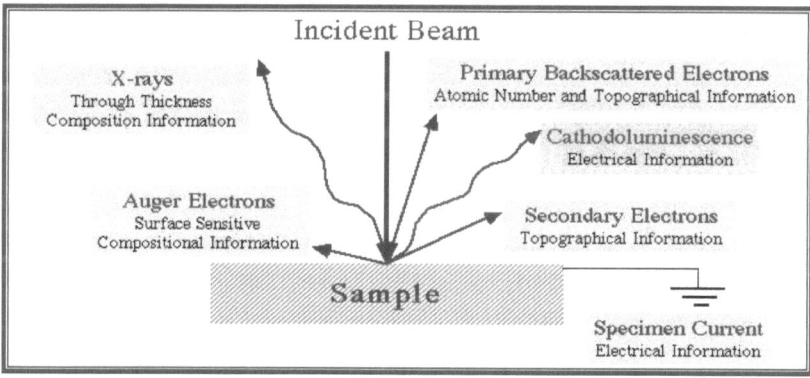

Fig. 2.1 Schematic diagram of the various signals generated when electron beam strikes the sample

SEM forms an image by rastering a highly focused electron beam, typically with energies of 1 to 20 KeV, across a sample and detecting the secondary or back scattered electrons ejected. The secondary electron originated from the top 5-15 nm of the sample and provides information on topography and to, a lesser extent on the elemental variation in the samples. Backscattered electrons escape from deeper with in the sample and provide information chiefly on the average atomic number of the sample.

The incident electron beam, when imparting energy to the sample, can cause emission of x-rays that are characteristics of the sample atoms. The energies of the x-ray are characterized in an energy dispersive spectrometer and can be used to identify elements in the samples. EDAX preliminary a qualitative technique and provides data from about a micrometer in diameter and micrometer deep.

Magnification in a SEM can be controlled over a range of up to 6 orders of magnitude from about 10 to 500,000 times. Unlike optical and transmission electron microscopes, image magnification in the SEM is not a function of the power of the objective lens. SEMs may have condenser and objective lenses, but their function is to focus the beam to a spot, and not to image the specimen. Provided the electron gun can generate a beam with sufficiently small diameter, a SEM could in principle work entirely without condenser or objective lenses, although it might not be very versatile or achieve very high resolution. In a SEM, as in scanning probe microscopy, magnification results from the ratio of the dimensions of the raster on the specimen and the raster on the display device. Assuming that the display screen has a fixed size, higher magnification results from reducing the size of the raster on the specimen, and vice versa. Magnification is therefore controlled by the current supplied to the x, y scanning coils, or the voltage supplied to the x, y deflector plates, and not by objective lens power. Typical applications of SEM are, in materials research, quality control, failure analysis, and forensic science. Industries that commonly use this technique include: semi-conductor and electronics, metals, ceramics, minerals, manufacturing, engineering, nuclear, paper, petroleum, bio-science and the motor industry.

2.6. Energy Dispersive X-ray microanalysis

Energy Dispersive X-Ray Analysis (EDX), also referred to as EDS and EDAX, is an x-ray technique used to identify the elemental composition of a specimen. Applications include materials and product research, troubleshooting, deformulation and more.EDX systems are attachments to SEM and TEM instruments where the imaging capability of the microscope is used to identify the specimen of interest. The data generated by EDX analysis consist of spectra showing peaks corresponding to the elements making up the true composition of the sample being analyzed. In a multi-technique approach EDX becomes very powerful, particularly in contamination analysis and industrial forensic science investigations. The technique can be qualitative, semi-quantitative, quantitative and also provide spatial distribution of elements through mapping.

The EDAX analysis system works as an integrated feature of a scanning electron microscope (SEM), and can not operate on its own. During EDAX Analysis, the specimen is strikes by an electron beam inside the scanning electron microscope. The electrons collide with the specimen atoms' own electrons, knocking some of them off in the process. A position vacated by an ejected inner shell electron is eventually occupied by a higher-energy electron from an outer shell. To be able to do so, however, the transferring outer electron must give up some of its energy by emitting an x-ray. The amount of energy released by the transferring electron depends on which shell it is transferring from, as well as which shell it is transferring to. Furthermore, the atom of every element releases x-rays with unique amounts of energy during the transferring process. Thus, by measuring the amounts of energy present in the x-rays being released by a specimen during electron beam bombardment, the identity of the atom from which the x-ray was emitted can be established. The output of an EDAX analysis is an EDAX spectrum. The EDAX spectrum is just a plot of how frequently an x-ray is received for each energy level. An EDAX spectrum normally displays peaks corresponding to the energy levels for which the most x-rays had been received. Each of these peaks is unique to an atom, and therefore corresponds to a single element. The higher a peak in a spectrum, the more concentrated the element is in the specimen.

2.7. Impedance Spectroscopy

Impedance spectroscopy, measures the dielectric properties of a medium as a function of frequency. It is based on the interaction of an external field with the electric dipole moment of the sample, often expressed by permittivity. This technique measures the impedance of a system over a range of frequencies, and therefore the frequency response of the system, including the energy storage and dissipation properties, is revealed.

"Impedance is known as the measure of the total opposition to flow of a sinusoidal electric current when using AC circuitry. To better understand what impedance actually is, one can use Ohms law to explain how it's measured. Ohm's law applies directly to simple resistors (and conductors) in both DC and AC circuits. However, in more complex AC circuits, the form of the current-voltage relationship defined by Ohms law must be changed from I=V/R to I = V/Z. Note that V is voltage, I is the resulting current, and R is resistance. The quantity Z for the modified formula is denoted as impedance, which is measured in ohms, and for a pure resistor Z = R. Impedance is commonly represented as $Z = R + iX$, where R is the ohmic resistance and X is the reactance".

Circuit elements such as inductors and capacitors have a frequency dependent opposition to current flow and increase or decrease current opposition as AC frequency changes. The opposition to current flow caused by inductance or capacitance is known as *reactance*. The term reactance refers to the imaginary part of the impedance measurement (iX) due to it being a function of the frequency. Generally speaking, a capacitor would have an impedance measurement that would decrease with increasing frequency supplied to the circuit. An inductor would have an impedance value that increases with increasing frequency. Also, a pure resistor would have the same impedance measurement at any frequency applied to the circuit. When capacitors or inductors are involved in an AC circuit, the current and voltage do not peak at the same time. The period difference between the peaks, expressed in degrees, is known as the phase difference. The contributions of capacitors and inductors differ in phase from resistive components by 90 degrees. For inductive circuits, this results in a positive phase because current lags behind the voltage while for a capacitive circuit, the current leads the voltage for a negative phase. A resistor does not cause a phase shift. Since the C/V behavior of capacitors and inductors are 180 degree out-of-phase, the impedance in a given circuit using both elements will have phase-angle dependence [49].

The application of alternating current (a.c.) method has been boosted by the fast development of microelectronics and computers. The a. c. method is based on treatment of a small (normally 10-50 mV<kT/e) a.c. signal, shifted or responded by impedance of the sample under investigation. Processes in the cell, electrode/ion conductor/electrode, involving charge carriers, is normally described in terms of finite elements such as resistance R (inverse electrolyte conductivity, Faradiac charge transfer), capacitance C (dielectric response of the material, double layer capacity), and in terms of complex elements: constant phase element (CPE) (capacity at fractral interface) and Warburg impedance W for diffusion at electrodes. Thus, at each frequency the cell has a certain complex impedance Z (ω), which can be expressed in series or a parallel connection of capacity and resistance. In a circuit, that contains resistance and impedance in a series the total impedance of the circuit is given by

$Z = R + (1/j\omega) C = R - (j/\omega C)$

Where $j = \sqrt{-1}$, ω is the angular frequency ($\omega = 2\pi f$) and R, C are not constant. This impedance contains real and imaginary terms (i. e. R and $1/j\omega C$ respectively) and is therefore called complex impedance Z*, where $Z^* = Z' - jZ''$. For a circuit which has R and C in parallel the complex impedance are given by the following equation: $Z' = R / 1+(\omega RC)^2$ and $Z'' = R. \omega RC / 1+(\omega RC)^2$. The complex plane representation is convenient to observe and graphically determine R, C values if the processes in the cell have sufficient different relaxation times (RC). In particular, a simple parallel connection of R and C (for bulk of an ideally homogenous material) yields a semicircle in a certain frequency range. Several impedance circuits have been used to describe the process in the cell: electrode/ solid ion conductor. Analysis of a.c. data is often carried out by complex plane methods which involve plotting the imaginary part of Z*, i.e. Z" against the real part Z'.

While investigating conductivity one should always bear in mind that the polycrystalline conductor consists of many grains and a thin grain boundary area (space). It may cause the appearance of two semicircles in the impedance spectrum described by the model. Ideally, the first circle at high frequency is related to the bulk properties (R bulk, C bulk in parallel) of the conductor, whereas second semicircle (R grain boundary, C grain boundary in parallel) observed at lower frequencies corresponds to the grain boundary in the ceramic. The sequence of both semicircles is defined by their relaxation times RC. Normally solid state conductors have a rather small dielectric constant of 1-20. Thus, low capacities of 10^{-11}-10^{-12} F/cm determine the fast relaxation of the bulk processes. A characteristic feature of the grain boundary effect is that the thin grain boundary area causes Cgb of 10^{-9} – 10^{-10} F/cm placing at lower frequencies. The resistance Rgb may have temperature characteristics different from Rb. Often, the grain boundary impedance is observed to be less apparent (limiting) in dense (about 100%) and very pure ceramics. However, at solid ionic materials, the ideal RC combination can be used rarely. Often impedance semicircles have their center below the real axis caused by time constant dispersion and capacities have to be replaced by the constant phase elements.

$Z(\omega) = 1/Q\,(i\omega)^n$ (where typically n= 0.5-1)

Each QR branch reflects a variation of impedance parameters with frequency. The origin of this effect for bulk and grain boundary phase differs from cases where the constant phase element is used for electrode impedance, for instance to describe Warburg impedance (n=0.5).

Impedance spectroscopy (IS) can provide accurate evaluation of material characteristics by measuring impedance values for a range of circuit elements whose capacitance or inductance characteristics depend in some way on their chemical environment.

2.8. FTIR Spectroscopy

Infrared (IR) spectroscopy is one of the most common spectroscopic techniques used by organic and inorganic chemists. Simply, it is the absorption measurement of different IR frequencies by a sample positioned in the path of an IR beam. The main goal of IR spectroscopic analysis is to determine the chemical functional groups in the sample. Different functional groups absorb characteristic frequencies of IR radiation. Using various sampling accessories, IR spectrometers can accept a wide range of sample types such as gases, liquids, and solids. Thus, IR spectroscopy is an important and popular tool for structural elucidation and compound identification [50]. An infrared spectrum represents a fingerprint of a sample with absorption peaks which correspond to the frequencies of vibrations between the bonds of the atoms making up the material. Because each different material is a unique combination of atoms, no two compounds produce the exact same infrared spectrum; therefore, infrared spectroscopy can result in a positive identification (qualitative analysis) of every different kind of material. In addition, the size of the peaks in the spectrum is a direct indication of the amount of material present. With modern software algorithms, infrared is an excellent tool for quantitative analysis.

Fourier Transform Infrared (FT-IR) spectrometry was developed in order to overcome the limitations encountered with dispersive instruments. The main difficulty was the slow scanning process. A method for measuring all of the infrared frequencies simultaneously, rather than individually, was needed. A solution was developed which employed a very simple optical device called an interferometer. The interferometer produces a unique type of signal which has all of the infrared frequencies "encoded" into it. The signal can be measured very quickly, usually on the order of one second or so. Thus, the time element per sample is reduced to a matter of a few seconds rather than several minutes.

IR spectrum of every molecule is unique; one of the most positive identification methods of an organic compound is to find a reference IR spectrum that matches that of the unknown compound. A large number of reference spectra for vapor and condensed phases are available in printed and electronic formats. The spectral libraries compiled by Sadtler and Aldrich are some of the most popular collections. In addition, spectral databases are often compiled according to application areas such as forensics, biochemicals and polymers. Computerized search programs can facilitate the matching process. In many cases where exact match to the spectrum of an unknown material cannot be found, these programs usually list the reference compounds that match the unknown spectrum most closely. This information is useful in narrowing the search. When it is combined with the data from other analysis such as NMR or mass spectrometry, a positive identification or high-confidence level tentative identification can often be achieved.

References

[1] Yoshio Waseda, Eiichiro Matsubara, Kozo Shinoda, X-Ray Diffraction Crystallography: Introduction, Examples and Solved Problems. Springer-Verlag, Berlin Heidelberg, **(2011)**.

[2] Jinghua Guo, X-Rays in Nanoscience: Spectroscopy, Spectromicroscopy and Scattering, John Wiley & Sons, **(2010)**.

[3] Jenny Pickworth Glusker, Kenneth N. Trueblood, Crystal Structure Analysis: A Primer, Oxford science publications, **(2010)**.

[4] A. K. Cheetham, A. P. Wilkinson, Angew. Chemie Int.Ed.Engl.31 **(1992)**: 1557.

[5] K. D. M. Harris, Tremayne, Chern. Mater.8 **(1996)**: 2554.

[6] J. I. Langford, D. Louër, Rep. Progr. Phys., 59 **(1996)**: 131.

[7] D. M. Poojary, A. Clearfield, Acc. Chern. Res. 30 **(1997)**: 414.

[8] A. Meden, Croat. Chern. Acta, 71 **(1998)**: 615.

[9] K. D. M. Harris, M. Tremayne, B. M. Kariuki, Angew. Chernie Int. Ed. 40 **(2001)**: 1626.

[10] H. M. Rietveld, J. Appl. Crystallogr. 2 **(1969)**: 65.

[11] R. A. Young (Editor), The Rietveld Method, International Union of Crystallography and Oxford University Press, Oxford, **(1993)**.

[12] J. W. Visser, J. Appl. Crystallogr. 2 **(1969)**: 89.

[13] P.-E. Werner, L. Eriksson, M. Westdahl, J. Appl.Crystallogr. 18 **(1985)**: 367.

[14] A. Boultif, D. Louër, J. Appl. Crystallogr. 24 **(1991)**: 987.

[15] R. A. Shirley, CRYSFIRE Suite of Programs for Indexing Powder Diffraction Patterns, University of Surrey.

[16] B. M. Kariuki, S. A. Belmonte, M. I. McMahon, R. L. Johnston, K. D. M. Harris, R. J. Nelmes, J. Synchrotron Radiation, 6 **(1999)**: 87.

[17] Artem R. Oganov, Modern Methods of Crystal Structure Prediction, John Wiley & Sons, **(2011)**.

[18] B. M. Kariuki, D. M. S. Zin, M. Tremayne, K. D. M.Harris, Chern. Mater. 8 **(1996)**: 565.

[19] M. Tremayne, B. M. Kariuki, K. D. M. Harris, J. Appl. Crystallogr. 29 **(1996)**: 211.

[20] M. Tremayne, B. M. Kariuki, K. D. M. Harris, J. Mat. Chem. 6 **(1996)**: 1601.

[21] Tremayne, B. M. Kariuki, K. D. M. Harris, Angew. Chemie Int. Ed. Engl. 36 **(1997)**: 770.

[22] K. D. M. Harris, B. M. Kariuki, M. Tremayne, Mat. Sci. Forum, 32 **(1998)**: 278.

[23] M. Tremayne, E. J. MacLean, C. C. Tang, C. Glidewell, Acta Crystallogr. 855 **(1999)**: 1068.

[24] J. M. Newsam, M. W. Deem, C. M. Freeman, Accuracy in Powder Diffraction II: NIST Special Publ. No. 846 **(1992)**: 80.

[25] D. Ramprasad, G. B. Pez, B. H. Toby, T. J. Markley, R. M. Pearlstein, J. Am. Chern. Soc. 117 **(1995)**: 10694.

[26] Y. G. Andreev, P. Lightfoot, P. G. Bruce, Chern. Commun. **(1996)**: 2169.

[27] C. M. Freeman, A. M. Gorman, J. M. Newsam, in Computer Modeling in Inorganic Crystallography (Editor: C.R.A. Catlow), Academic Press, San Diego, **(1997)**.

[28] Y. G. Andreev, P. Lightfoot, P. G. Bruce, J. Appl. Crystallogr.30 **(1997)**: 294.

[29] W. I. F. David, K. Shankland, N. Shankland, Chern. Commun. **(1998)**: 931.

[30] G. E. Engel, S. Wilke, O. Konig, K. D. M. Harris, F. J. J. Leusen, J. Appl. Crystallogr. 32 **(1999)**: 1169.

[31] B. M. Kariuki, H. Serrano-Gonzalez, R. L. Johnston, K. D. M. Harris, Chern. Phys. Lett. 280 **(1997)**: 189.

[32] K. D. M. Harris, R. L. Johnston, B. M. Kariuki, M. Tremayne, J. Chern. Res. (S) **(1998)**: 390.

[33] K. D. M. Harris, R. L. Johnston, B. M. Kariuki, Acta Crystallogr. 632 **(1998)**: A-54.

[34] K. D. M. Harris, R. L. Johnston, B. M. Kariuki, Anales de Química, Int. Ed., 94 **(1998)**: 410.

[35] G. W. Turner, E. Tedesco, K. D. M. Harris, R. L. Johnston, B. M. Kariuki, Chern. Phys. Lett. 183 **(2000)**: 321.

[36] B. M. Kariuki, P. Calcagno, K. D. M. Harris, D. Philp, R. L. Johnston, Angew. Chemie Int. Ed. 38 **(1999)**: 831.

[37] G. Reck, R.-G. Kretschmer, L. Kutschabsky, W. Pritzkow, Acta Crystallogr. 417 **(1988)**: A-44.

[38] N. Masciocchi, R. Bianchi, P. Cairati, G. Mezza, T. Pilati, A. Sironi, J. Appl. Crystallogr. 27 **(1994)**: 426.

[39] R. E. Dinnebier, P. W. Stephens, J. K. Carter, A. N.Lommen, P. A. Heiney, A. R. Mcghie, L. Brard, A. B. Smith III, J. Appl. Crystallogr. 28 **(1995)**: 327.

[40] R. B. Hammond, K. J. Roberts, R. Docherty, M.Edmondson, J. Phys. Chem. 6532 **(1997)**: 8101.

[41] Louer D, Louer M.J. Methode d'essais et erreurs pour l'indexation automatique des diagrammes de poudre. J Appl Cryst. 5 **(1972)**: 271.

[42] Werner P-E Trial-and-error computer methods for the indexing of unknown powder patterns. Z Kristallogr 120 **(1964)**: 375.

[43] Werner P-E On the determination of unit cell dimensions from inaccuratepowder diffraction data. J Appl Cryst 9 **(1976)**: 216.

[44] R. Shirley, Progress in Automatic Powder Indexing, http://www.ccp14. ac.uk/posteralks/ Shrley_powdind_epdic2000/index.htm **(May 19 2005)**.

[45] Coelho AA) Indexing of powder diffraction patterns by iterative use of singular value decomposition. J Appl Cryst 36 **(2003)**: 86.

[46] A. C. Larson, R. B. Von Dreele, General Structure Analysis System; Los Alamos National Laboratory: Los Alamos, NM, Report LAUR, **(2000)**: 86-748.

[47] B. H.Toby, EXPGUI, a graphical user interface for GSAS. J. Appl. Crystall., 34 **(2001)**: 210.

[48] International Union of Cryastallography (IUCr) workshop on Rietveld method, Netherlands 13-15 June 1989. Edited by R. A. Young, Oxford University Press. Inc. Newyork Reprinted: **(1996)**.

[49] Vadim F. Lvovich, Impedance Spectroscopy: Applications to Electrochemical and Dielectric Phenomena, John Wiley & Sons: **(2011)**.

[50] Brian C. Smith, Fundamentals of Fourier Transform Infrared Spectroscopy, Second Edition, CRC press: **(2011)**.

Chapter 3

Investigational Procedures

3.1 Characterization and synthesis of materials

Ceramic precursors of NZP, CZP and Perovskite family can be prepared by different synthetic routes (viz- Sol-gel, hydrothermal, solid state reaction etc.) and in various forms including single crystals, polycrystalline ceramics and thick or thin films. Polycrystalline samples can be prepared by solid state sintering, involving repeated steps of mixing, grinding and heating of oxides until a single-phase and high density material is obtained. Ceramic route method was developed by Australian workers who for the first time prepared the composite precursors of hollandites, zirconolites and perovskites of variables compositions. Titania and Zirconiua based ceramic samples under investigation in the present work were polycrystalline ceramics obtained by the solid state reaction technique.

A solid-state reaction is a direct reaction between starting reagents (usually powders) at high temperature. High temperature provides the necessary energy for the reaction to occur. Solid-state reaction is usually slow because during the reaction, a large amount of bonds break, and the ions migrate through a solid unlike gas phase and solution reactions. The limiting factor in solid state reaction is usually diffusion. So the rate controlling step in a solid-state reaction is the diffusion of the cations through the product layer. Solid-state reaction occurs much more quickly with increasing temperature and reaction does not normally occur until the reaction-temperature reaches at least $2/3^{rd}$ of the melting point of one of the reactants.

The objective of ceramic route synthesis of Titania and Zirconia based ceramic precursors is to understand the crystallochemical interaction of various cations with the host matrix and their effect on electrical properties of materials. This chapter gives a detailed description of the sample preparation, followed by a discussion on sample composition and morphology. The general procedure of synthesis can be summarized as follows:

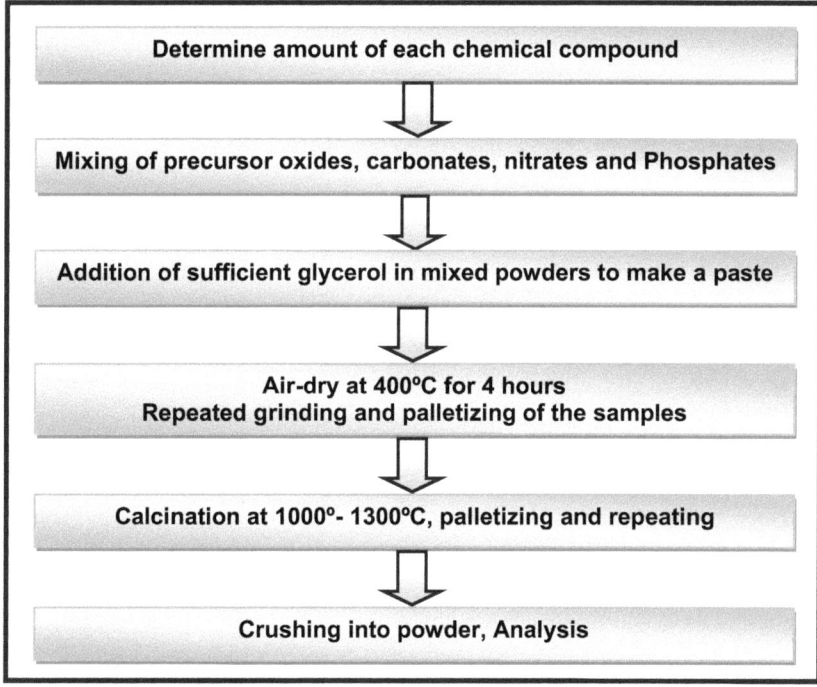

Synthesis of substituted sodium zirconium phosphate and calcium zirconium phosphate has been carried out as follows: the Stoichiometric quantities of sodium, calcium, zirconium, phosphorous and substituent cations corresponding to the molecular formula $Na_{1-x}M_{x/2}Zr_2P_3O_{12}$ / $Na_{1-x}(M'_{1.33}M''_1)_xZr_2P_3O_{12}$/ $Ca_{1-2x}Zr_4M_{2x}P_{6-2x}O_{24}$ were thoroughly grinded using mortar and pestle. AR quality Na_2CO_3, $Zr-O(NO_3)_2.xH_2O$, $NH_4Mo_7O_{24}.4H_2O$ (Ammonium molybdate), $Sr(NO_3)_2$, $CsNO_3$, $CaCO_3$ and $(NH_4)_6H_2PO_4$ were used in appropriate quantities as source of Na, Ca, Zr, P and M (Cs, Sr and Mo) respectively. The mixtures so prepared were heated on stainless steel tray initially at 400°C for 4 hours. This initial heating is done to decompose carbonates, nitrates and $(NH_4)H_2PO_4$ with emission of CO_2, NO_x, ammonia and water vapors. The decomposition losses were made up by adding the calculated quantities. The mixture was reground to micron size, pressed into pellets and sintered in a platinum crucible at 1050°C for 50 hours. The process was repeated to get a polycrystalline dense material. The following ceramic phases have been synthesized during the course of investigations:

Sodium zirconium phosphates (NZP):

- Sodium Zirconium Metal Phosphate $Na_{1-x}M_{x/2}Zr_2P_3O_{12}$ [M = Sr and x = 0.1-1.0]
- Co- substitution in Sodium Zirconium Metal Phosphate
  $Na_{1-x}(M'_{1.33}M''_1)_xZr_2P_3O_{12}$ [M'=Cs, M''= Sr and x = 0.1-1.0]

Calcium zirconium phosphates (CZP):

- Calcium Zirconium Metal Phosphate
  $Ca_{1-2x}Zr_4M_{2x}P_{6-2x}O_{24}$ [M = Mo and x = 0.1-0.5]

Perovskite:

- Metal Calcium Titanate
  $Ca_{1-x}M_xTiO_3$ [M = Sm and x = 0.1-0.5]

3.2 Characterization of ceramic materials

3.2.1 Analysis of phases using rietveld refinement

The synthetic ceramic materials have been characterized by powder X-ray diffraction analysis using Pan Analytical XPERT-PRO powder diffractometer. About 0.5gm of sample was required. Each polycrystalline sample was ground to a fine powder with a mortar and pestle before being pressed into the sample holder. CuKα (λ =1.54056) radiation was used. The Powder XRD patterns were recorded over a diffraction-angle (2θ) range of 10° to 120°, with a step size of 0.02° and a counting time of 5 seconds per step. The phase purity of each ceramic sample has been verified using standard pattern reported in JCPDS (Joint Committee on Powder Diffraction Standards) file. Powder diffraction step analysis data of each phase then subjected to Rietveld refinement using GSAS software.

3.2.2 Morphology and elemental analysis using SEM/EDAX

The grain size and surface morphology of the Samples have been investigated by scanning electron microscopy (SEM) using a HITACHI S-3400N electron microscope equipped with 'Thermonoran' ultra dry detector facility for energy dispersive X-ray (EDAX) analysis. A computer based SEM image analysis was carried out to estimate the grain size of the samples. For each sample, the size of grains (chosen at random) was measured. The elemental composition and homogeneity of the synthesized samples were investigated by means of quantitative energy dispersive X-ray spectroscopy (EDS). It is one of the most versatile and widely used tools of modern science as it allows the study of both morphology and composition of materials. By scanning an electron probe across a specimen, high-resolution images of the morphology or topology of a specimen, with great depth of field, at very low or very high magnifications can be obtained. Compositional analysis of a material may also be obtained by monitoring secondary X-ray produced by the electron specimen interaction. Thus, detailed maps of elemental distribution can be produced from materials. Characterization of fine particulate matter in terms of size, shape and distribution as well as statistical analyses of these parameters has been performed. The SEM used in the ceramic study is of routine morphological nature combined with energy dispersive (EDAX) analysis to obtain elemental profile in the crystal & bulk phases.

3.3 Electrical properties Study

Ceramic compounds have been investigated by a.c. impedance spectroscopy carried out on impedance analyzer HIOKI LCR HITESTER 3532-50 between the frequency ranges from 42Hz to 5MHz. Impedance, capacitance and dissipation factor were collected as a function of frequency in the range of ambient temperature to 300°C over heating and cooling cycles. The sample in the form of circular disc (1.29 cm diameter and 0.30 cm thickness) was used for measurements. Silver paint applied to both surfaces of the pellet served as electrodes. Temperature dependence of dielectric constant ε and dielectric loss (tanδ) were measured to determine the transition temperature (Tc) at different frequencies.

The temperature and frequency dependent electrical properties i.e., complex permittivity (ε^*), complex impedance (Z^*), complex electrical module (M^*) and electrical loss or dissipation factor (tan∂) can be written as:

$$Z^* = Z' - jZ'' = 1/jC_0 \varepsilon^* \omega, \quad \varepsilon^* = \varepsilon' - j\varepsilon'',$$
$$M^* = M' + jM'' = 1/\varepsilon^* = j\omega \varepsilon_0 Z^*$$

and

$$\tan\partial = \varepsilon''/\varepsilon' = M''/M' = Z'/Z'' = Y'/Y''$$

where $\omega = 2\pi f$ is the angular frequency (f = frequency), C_0 is the geometrical capacitance, $j = \sqrt{-1}$ and the symbols 'and' are real and imaginary parts of the above complex parameters. these relation offers a wide scope for a graphical analysis of the various parameters under different experimental conditions (temperature and frequency).

3.3.1 Dielectric measurements

For detailed information on electrical behavior, dielectric measurements are made over the range of frequencies covering audio frequency and radiofrequency regions. The results are plotted as frequency vs. dielectric loss factor (tan delta), temperature vs. loss factor, frequency vs. dielectric constant and temperature vs. dielectric constant. The dielectric loss factor is given by the ratio $\varepsilon''/\varepsilon'$.

Where $\varepsilon"$ is the total loss factor and is the measure of conductance in the material.

3.4 FTIR Spectroscopy

To confirm functional compositions of the phosphates, their IR spectra were recorded SHIMADZU FTIR-8400S instrument. Samples were prepared by finely dispersing powder material on a KBr carrier.

3.5 Experimental data

Table 3.1 Crystallographic data for $Na_{1-x}Sr_{x/2}Zr_2P_3O_{12}$ (x = 0.1-1.0) at room temperature

Structure	Rhombohedral	
Space group	R-3c	
Z	6	
$\alpha = \beta$	=	90°
γ	=	120°

Parameters	$Na_{0.9}Sr_{0.05}Zr_2P_3O_{12}$	$Na_{0.5}Sr_{0.25}Zr_2P_3O_{12}$	$Sr_{0.5}Zr_2P_3O_{12}$
Lattice constants			
a = b	8.81716(12)	8.82886(21)	8.80736(13)
c	22.7741(5)	22.8080(7)	22.9307(6)
Rp	0.0998	0.0980	0.0990
Rwp	0.1295	0.1292	0.1292
R expected	0.0877	0.0770	0.0758
RF^2	0.12572	0.16783	0.17312
Volume of unit cell	1533.31(4)	1539.67(9)	1540.42(4)
S (GoF)	2.179	2.819	2.908
DWd	0.990	0.732	0.738
Unit cell formula weight	2954.532	3004.500	3066.960

Density$_{X\text{-ray}}$ (gm/cm^3)	3.200	3.240	3.306
Slope	1.34	1.5171	1.4879

Table 3.2 Refined atomic coordinates of $Na_{1-x}Sr_{x/2}Zr_2P_3O_{12}$ (x = 0.1-1.0) polycrystalline solid solution at room temperature

Atom	x	y	z	Occupancy	Uiso (Å^2)
$Na_{0.9}Sr_{0.05}Zr_2P_3O_{12}$					
Na	0.0	0.0	0.0	0.9	0.04731
Sr	0.0	0.0	0.0	0.05	0.28517
Zr	0.0	0.0	0.14569	1.0	0.012
P	0.29347	0.0	0.25	1.0	0.0189
O1	0.17594	-0.02425	0.19628	1.0	0.03243
O2	0.194	0.17051	0.08888	1.0	0.03243
$Na_{0.7}Sr_{0.15}Zr_2P_3O_{12}$					
Na	0.0	0.0	0.0	0.7	0.11265
Sr	0.0	0.0	0.0	0.15	0.06081
Zr	0.0	0.0	0.14569	1.0	0.03041
P	0.29347	0.0	0.25	1.0	0.03041
O1	0.17594	-0.02425	0.19628	1.0	0.03177
O2	0.194	0.17051	0.08888	1.0	0.03177

$Na_{0.5}Sr_{0.25}Zr_2P_3O_{12}$

Na	0.0	0.0	0.0	0.5	0.8
Sr	0.0	0.0	0.0	0.25	0.09986
Zr	0.0	0.0	0.14569	1.0	0.07708
P	0.29347	0.0	0.25	1.0	0.07708
O1	0.17594	-0.02425	0.19628	1.0	0.08668
O2	0.194	0.17051	0.08888	1.0	0.08668

$Na_{0.3}Sr_{0.35}Zr_2P_3O_{12}$

Na	0.0	0.0	0.0	0.3	0.8
Sr	0.0	0.0	0.0	0.35	0.11956
Zr	0.0	0.0	0.14569	1.0	0.03536
P	0.29347	0.0	0.25	1.0	0.03536
O1	0.17594	-0.02425	0.19628	1.0	0.04111
O2	0.194	0.17051	0.08888	1.0	0.04111

$Sr_{0.5}Zr_2P_3O_{12}$

Sr	0.0	0.0	0.0	0.5	0.17684
Zr	0.0	0.0	0.14569	1.0	0.06634
P	0.29347	0.0	0.25	1.0	0.06634
O1	0.17594	-0.02425	0.19628	1.0	0.08286
O2	0.194	0.17051	0.08888	1.0	0.08286

Table 3.3 Inter atomic distances (Å) and polyhedral distortion of polycrystalline $Na_{1-x}Sr_{x/2}Zr_2P_3O_{12}$ (x = 0.1-1.0) ceramic phase

Bond lengths(Å) / Distortion (Δ)	$Na_{0.9}Sr_{0.05}Zr_2P_3O_{12}$	$Na_{0.7}Sr_{0.15}Zr_2P_3O_{12}$	$Na_{0.5}Sr_{0.25}Zr_2P_3O_{12}$	$Na_{0.3}Sr_{0.35}Zr_2P_3O_{12}$	$Sr_{0.5}Zr_2P_3O_{12}$
Na–O2	2.59071(4)*6	2.58940(4)*6	2.59440(6)*6	2.590890(30)*6	2.60048(5)*6
Zr–O1	2.027640(20)*3	2.026540(30)*3	2.03044(4)*3	2.027720(20)*3	2.030630(30)*3
Zr–O2	2.070850(20)*3	2.069750(30)*3	2.07373(4)*3	2.070950(20)*3	2.075020(30)*3
P–O1	1.547510(20)*2	1.546730(30)*2	1.549720(30)*2	1.547620(20)*2	1.553530(30)*2
P–O2	1.518260(20)*2	1.517370(30*2	1.52028(4)*2	1.518260(20)*2	1.516660(20)*2
ZrO_6 (Δ ×10^4)	2.2778	2.2790	2.2830	2.2798	2.3997
PO_4 (Δ × 10^4)	1.3953	1.4066	1.4115	1.4058	2.2138

- The * Denotes multiplicity of bonds and angles
- The values in parentheses denotes esd (estimated standard deviation) values

Table 3.4 O-M-O bond angles of polycrystalline $Na_{1-x}Sr_{x/2}Zr_2P_3O_{12}$ (x = 0.1-1.0) ceramic phases

O-M-O Bond angle(deg.)	$Na_{0.9}Sr_{0.05}Zr_2P_3O_{12}$	$Na_{0.7}Sr_{0.15}Zr_2P_3O_{12}$	$Na_{0.5}Sr_{0.25}Zr_2P_3O_{12}$	$Na_{0.3}Sr_{0.35}Zr_2P_3O_{12}$	$Sr_{0.5}Zr_2P_3O_{12}$
O2–Na–O2	65.4377(12)*6	65.4317(14)*6	65.4304(18)*6	65.4325(10)*6	65.0799(14)*6
O2–Na–O2	180.0(0)*2	179.972(3)*2	180.0(0)*2	179.980(2)*2	180.0(0)*2
O2–Na–O2	180.0(0)	180.0(0)	179.972	180.0(0)	179.9657
O2–Na–O2	114.5623(12)*6	114.5683(14)*6	114.5696(18)*6	114.5675(10)*6	114.9201(14)*6
O1–Zr–O1	90.8993 (10)*3	90.8942 (12)*3	90.8932(15)*3	90.8950(8)*3	90.5992(12)*3
O1–Zr–O2	91.9354(9)*3	91.9406(12)*3	91.9417(15)*3	91.9399(9)*3	92.2492(12)*3
O1–Zr–O2	175.9612(0)*3	175.9608(0)*3	175.961(0)*3	175.961(0)*3	175.9521(0)*3
O1–Zr–O2	91.9321(10)*3	91.9374(12)*3	91.9385(15)*3	91.9367(9)*3	92.2460(12)*3
O2–Zr–O2	85.0955(11)*3	85.0900(13)*3	85.0889(16)*3	85.0908(9)*3	84.7679(13)*3
O1–P–O1	106.1800(14)	106.1872(17)	106.1886(21)	106.1861(12)	106.6079(17)
O1–P–O2	107.98900(10)*2	107.98850(10)*2	107.98840(20)*2	107.98850(10)*2	107.95860(10)*2
O1–P–O2	111.9165(6)*2	111.9134(7)*2	111.9128(9)*2	111.9138(5)*2	111.7356(7)*2
O2–P–O2	110.7953(0)	110.7954(0)	110.7954(0)	110.7954(0)	110.7997(0)

Table 3.5 Observed and calculated structure factors of polycrystalline $Na_{1-x}Sr_{x/2}Zr_2P_3O_{12}$ (x = 0.1-1.0) ceramic phase. The seven columns within each group contain the values h, k, l, d-spacing, structure factor, Fosq (observed), Fcsq (calculated) and intensity respectively. The reflection selected from the CIF output of the final cycle of the refinement.

h	k	l	d-Space	F^2 (Obs.)	F^2 (Calc.)	Intensity (%)
$Na_{0.9}Sr_{0.05}Zr_2P_3O_{12}$						
1	0	-2	6.34197	2.717 E+04	2.632 E+04	20.7199
1	0	-2	6.34197	2.762 E+04	2.632 E+04	10.4855
1	0	4	4.56438	1.493 E+05	1.448 E+05	62.5194
1	0	4	4.56438	1.494 E+05	1.448 E+05	31.1393
1	1	0	4.40858	1.836 E+05	1.818 E+05	72.3306
1	1	0	4.40858	1.872 E+05	1.818 E+05	36.7240
1	1	3	3.81234	1.299 E+05	1.190 E+05	79.6940
1	1	3	3.81234	1.269 E+05	1.190 E+05	38.7778
2	0	-4	3.17099	2.317 E+05	1.995 E+05	52.6320
2	0	-4	3.17099	2.227 E+05	1.995 E+05	25.1873
1	1	6	2.87644	2.560 E+05	2.142 E+05	99.9998
1	1	6	2.87644	2.373 E+05	2.142 E+05	46.1762
2	1	1	2.86319	2.297 E+04	2.268 E+04	8.9118
2	1	4	2.57425	8.547 E+04	6.232 E+04	28.3360
2	1	4	2.57425	8.962 E+04	6.232 E+04	14.8048
3	0	0	2.54530	2.497 E+05	2.372 E+05	40.7275
3	0	0	2.54530	2.265 E+05	2.372 E+05	18.4059
2	0	8	2.28219	6.020 E+04	5.031 E+04	8.4383
2	2	0	2.20429	3.858 E+04	3.570 E+04	5.1632
1	1	9	2.19463	2.540 E+04	2.412 E+04	6.7585

3	0	-6	2.11399	8.290 E+04	6.243 E+04	10.5033
2	1	-8	2.02673	1.085 E+05	9.482 E+04	26.0602
2	1	-8	2.02673	9.401 E+05	9.482 E+04	11.2540
3	1	-4	1.98494	6.875 E+04	6.858 E+04	16.0860
3	1	-4	1.98494	6.145 E+04	6.858 E+04	7.1672
2	0	-10	1.95587	1.217 E+05	1.017 E+05	13.9770
2	0	-10	1.95587	1.470 E+05	1.017 E+05	8.4191
2	2	6	1.90617	1.900 E+05	1.643 E+05	42.2950
2	2	6	1.90617	1.675 E+05	1.643 E+05	18.5888
2	1	10	1.78783	1.778 E+05	1.682 E+05	36.6816
2	1	10	1.78783	1.600 E+05	1.682 E+05	16.4556
3	1	8	1.69918	8.567 E+04	7.382 E+04	16.6666
3	1	8	1.69918	7.528 E+04	7.382 E+04	7.3026
3	2	4	1.67433	1.048 E+05	9.227 E+04	20.0440
3	2	4	1.67433	1.057 E+05	9.227 E+04	10.0841
4	1	0	1.66629	1.921 E+05	1.739 E+05	36.5630
4	1	0	1.66629	1.732 E+05	1.739 E+05	16.4377
1	0	-14	1.59102	1.261 E+05	1.179 E+05	11.4036
1	0	-14	1.59102	1.246 E+05	1.179 E+05	5.6192
4	0	-8	1.58549	8.109 E+04	7.444 E+04	7.3033
3	1	-10	1.55087	1.463 E+05	1.349 E+05	25.7352
3	1	-10	1.55087	1.350 E+05	1.349 E+05	11.8424
4	1	6	1.52574	7.898 E+04	7.197 E+04	13.6473
4	1	-6	1.52574	8.126 E+04	7.197 E+04	14.0413
4	1	-6	1.52574	7.234 E+04	7.197 E+04	6.2334
4	1	6	1.52574	7.033 E+04	7.197 E+04	6.0601

2	0	14	1.49654	1.908 E+05	1.859 E+04	16.1455
2	0	14	1.49654	1.740 E+05	1.859 E+04	7.3461
5	0	-4	1.47504	1.083 E+05	1.020 E+05	9.0308
3	3	0	1.46953	1.143 E+05	1.225 E+05	9.4868
4	0	10	1.46299	9.762 E+04	1.119 E+05	8.0670
1	1	15	1.43553	3.389 E+04	2.971 E+04	5.4909
2	1	-14	1.41712	1.067 E+05	1.048 E+05	17.0639
2	1	-14	1.41712	9.289 E+04	1.048 E+05	7.4068
4	2	-4	1.39882	4.968 E+04	4.740 E+04	7.8364
3	2	10	1.38853	5.790 E+04	6.022 E+04	9.0646
3	3	6	1.37041	3.926 E+04	2.890 E+04	6.0645
5	1	4	1.33331	1.078 E+05	1.021 E+05	16.1850
5	1	4	1.33331	9.610 E+04	1.021 E+05	7.1990
3	1	14	1.29007	9.175 E+04	8.824 E+04	13.3295
3	1	14	1.29007	8.144 E+04	8.824 E+04	5.9012
6	0	0	1.27265	2.019 E+05	2.019 E+05	14.4685
6	0	0	1.27265	1.820 E+05	2.019 E+05	6.5029
0	0	18	1.26523	2.263 E+05	2.477 E+05	5.3728
4	2	-10	1.21895	4.624 E+04	5.028 E+04	6.3479
3	2	-14	1.19203	1.116 E+05	1.061 E+05	14.9799
3	2	-14	1.19203	1.022 E+05	1.061 E+05	6.8401
5	2	-6	1.16382	4.031 E+04	3.946 E+04	5.2848
5	2	6	1.16382	6.544 E+04	3.946 E+04	8.5781
4	3	-8	1.14861	6.908 E+04	5.988 E+04	8.9378
1	0	-20	1.12625	9.996 E+04	9.844 E+04	6.3413
5	0	14	1.11341	1.158 E+05	8.999 E+04	7.2615

4	3	10	1.09938	1.171 E+05	1.097 E+05	14.5062
4	3	10	1.09938	1.216 E+05	1.097 E+05	7.5094
4	2	14	1.07951	6.771 E+04	5.826 E+04	8.2353
5	1	-14	1.04853	9.052 E+04	9.053 E+04	10.6925
5	1	-14	1.04853	9.927 E+04	9.053 E+04	5.8480
6	2	4	1.04105	6.881 E+04	6.462 E+04	8.0692
6	1	-10	1.03679	5.831 E+04	4.788 E+04	6.8100
7	1	0	1.01140	4.550 E+04	4.096 E+04	5.1827
3	1	20	1.00292	6.362 E+04	5.303 E+04	7.1858
7	0	10	0.98381	1.101 E+05	8.886 E+04	6.1000
5	3	-10	0.98381	5.329 E+04	4.305 E+04	5.9036
4	0	-20	0.97794	1.050 E+05	8.552 E+04	5.7827
5	4	4	0.96357	6.221 E+04	5.081 E+04	6.7510
1	1	24	0.92767	6.366 E+04	6.545 E+04	6.6574
5	3	14	0.90600	5.594 E+04	3.671 E+04	5.7255
$Na_{0.7}Sr_{0.15}Zr_2P_3O_{12}$						
1	0	-2	6.33852	3.812 E+04	3.733 E+04	17.1558
1	0	-2	6.33852	3.732 E+04	3.733 E+04	8.3585
1	0	4	4.56210	2.368 E+05	2.300 E+05	58.6764
1	0	4	4.56210	2.481 E+05	2.300 E+05	30.6032
1	1	0	4.40600	2.613 E+05	2.826 E+05	60.9149
1	1	0	4.40600	2.596 E+05	2.826 E+05	30.1321
1	1	3	3.81023	1.981 E+05	1.883 E+05	72.0915
1	1	3	3.81023	1.879 E+05	1.883 E+05	34.0422
2	0	-4	3.16926	3.120 E+05	2.942 E+05	42.1414
2	0	-4	3.16926	2.948 E+05	2.942 E+05	19.8344

1	1	6	2.87498	4.297 E+05	3.491 E+05	99.9999
1	1	6	2.87498	3.771 E+05	3.491 E+05	43.7098
2	1	1	2.86152	3.290 E+04	3.423 E+04	7.6031
2	1	4	2.57281	1.463 E+05	8.073 E+04	28.9634
2	1	4	2.57281	1.258 E+05	8.073 E+04	12.4030
3	0	0	2.54380	3.400 E+05	3.330 E+05	33.1165
2	0	8	2.28105	7.754 E+04	8.064 E+04	6.5051
1	1	9	2.19356	3.469 E+04	3.854 E+04	5.5286
3	0	-6	2.11284	1.052 E+05	9.316 E+05	7.9881
2	1	-8	2.02568	1.438 E+05	1.469 E+05	20.7276
2	1	-8	2.02568	1.344 E+05	1.469 E+05	9.6568
3	1	-4	1.98381	8.598 E+04	8.707 E+05	12.0757
3	1	-4	1.98381	8.368 E+04	8.707 E+05	5.8585
2	0	-10	1.95492	1.722 E+05	1.436 E+05	11.8808
2	0	-10	1.95492	2.130 E+05	1.436 E+05	7.3249
2	2	6	1.90512	2.687 E+05	2.305 E+05	35.9346
2	2	6	1.90512	2.317 E+05	2.305 E+05	15.4483
2	1	10	1.78693	2.343 E+05	2.396 E+05	29.0725
2	1	10	1.78693	2.066 E+05	2.396 E+05	12.7860
3	1	8	1.69827	1.362 E+05	1.005 E+05	15.9604
3	1	8	1.69827	1.146 E+05	1.005 E+05	6.6933
3	2	4	1.67337	1.288 E+05	1.092 E+05	14.8436
3	2	4	1.67337	1.276 E+05	1.092 E+05	7.3348
4	1	0	1.66531	2.376 E+05	2.239 E+05	27.2445
4	1	0	1.66531	2.295 E+05	2.239 E+05	13.1191
4	0	-8	1.58463	9.801 E+04	9.473 E+04	5.3237

3	1	-10	1.55006	2.108 E+05	1.702 E+05	22.3702
3	1	-10	1.55006	1.741 E+05	1.702 E+05	9.2103
4	1	6	1.52488	9.484 E+04	8.515 E+04	9.8912
4	1	6	1.52488	9.632 E+04	8.515 E+04	5.0098
4	1	-6	1.52488	9.852 E+04	8.784 E+04	10.2747
4	1	-6	1.52488	9.942 E+04	8.784 E+04	5.1709
2	0	14	1.49583	2.404 E+04	2.572 E+04	12.2854
2	0	14	1.49583	2.357 E+04	2.572 E+04	6.0072
5	0	-4	1.47418	1.175 E+05	1.184 E+05	5.9134
3	3	0	1.46867	1.221 E+04	1.342 E+04	6.1228
4	0	10	1.46221	1.104 E+05	1.284 E+05	5.5094
2	1	-14	1.41643	1.233 E+05	1.276 E+05	11.9177
2	1	-14	1.41643	1.257 E+05	1.276 E+05	6.0578
4	2	-4	1.39801	5.962 E+05	4.729 E+04	5.6847
3	2	10	1.38779	5.954 E+05	6.758 E+05	5.6353
5	1	14	1.33254	1.109 E+05	1.136 E+05	10.0733
3	1	14	1.28943	1.017 E+05	1.031 E+05	8.9378
6	0	0	1.27190	2.110 E+05	2.321 E+05	9.1526
3	2	-14	1.19142	1.156 E+05	1.118 E+05	9.3975
3	2	-14	1.19142	1.403 E+05	1.118 E+05	5.6898
5	2	6	1.16316	6.711 E+04	6.504 E+04	5.3292
4	3	-8	1.14797	7.164 E+05	6.266 E+05	5.6151
5	1	-14	1.04798	9.880 E+04	8.432 E+04	7.0707
$Na_{0.5}Sr_{0.25}Zr_2P_3O_{12}$						
1	0	-2	6.35070	3.131 E+03	2.524 E+03	7.8393
1	0	4	4.57091	1.744 E+04	1.512 E+04	39.7782

1	1	0	4.41443	1.831 E+04	1.862 E+04	41.4315
1	1	0	4.41443	1.854 E+04	1.862 E+04	20.9695
1	1	3	3.81755	1.424 E+04	1.195 E+04	62.6239
1	1	3	3.81755	1.314 E+04	1.195 E+04	28.8827
2	0	-4	3.17535	2.054 E+04	1.831 E+04	43.7817
2	0	-4	3.17535	1.934 E+04	1.831 E+04	20.6080
1	1	6	2.88052	2.382 E+04	1.916 E+04	100.000
1	1	6	2.88052	2.085 E+04	1.916 E+04	43.7402
2	1	1	2.86700	1.935 E+03	1.982 E+03	8.1194
2	1	4	2.57775	7.387 E+03	4.865 E+03	30.4671
2	1	4	2.57775	6.569 E+03	4.865 E+03	13.5405
3	0	0	2.54867	1.684 E+04	1.815 E+04	34.6695
3	0	0	2.54867	1.623 E+04	1.815 E+04	16.6963
2	0	8	2.28545	3.658 E+03	3.721 E+03	7.3891
1	1	9	2.19781	1.528 E+03	1.853 E+03	6.1281
3	0	-6	2.11690	4.641 E+03	4.633 E+03	9.2412
2	1	-8	2.02958	5.928 E+03	6.297 E+03	23.4097
2	1	-8	2.02958	5.673 E+03	6.297 E+03	11.1943
3	1	-4	1.98761	3.456 E+03	4.130 E+03	13.5886
3	1	-4	1.98761	3.484 E+03	4.130 E+03	6.8443
2	0	-10	1.95870	6.502 E+03	6.411 E+03	12.7414
2	0	-10	1.95870	7.813 E+03	6.411 E+03	7.6506
2	2	6	1.90878	1.029 E+04	9.791 E+03	40.0936
2	2	6	1.90878	9.115 E+03	9.791 E+03	17.7466
2	1	10	1.79038	8.543 E+03	9.449 E+03	32.7692
2	1	10	1.79038	7.484 E+03	9.449 E+03	14.3452

3	1	8	1.70153	4.920 E+03	3.749 E+03	18.6134
3	1	8	1.70153	4.073 E+03	3.749 E+03	7.7001
3	2	4	1.67658	4.256 E+03	4.256 E+03	16.0346
3	2	4	1.67658	4.165 E+03	4.256 E+03	7.8392
4	1	0	1.66850	7.867 E+03	8.212 E+03	29.5445
4	1	0	1.66850	7.549 E+03	8.212 E+03	14.1896
1	0	-14	1.59337	5.013 E+03	5.481 E+03	9.2943
1	0	-14	1.59337	5.144 E+03	5.481 E+03	9.2943
3	1	-10	1.55305	6.705 E+03	5.643 E+03	24.6485
3	1	-10	1.55305	5.556 E+03	5.643 E+03	10.2036
4	1	6	1.52781	3.057 E+03	2.857 E+03	11.1732
4	1	-6	1.52781	3.142 E+03	2.941 E+03	11.4844
4	1	-6	1.52781	3.163 E+03	2.941 E+03	5.7747
4	1	6	1.52781	3.080 E+03	2.857 E+03	5.6242
2	0	14	1.49873	7.530 E+03	7.774 E+03	13.6630
2	0	14	1.49873	7.309 E+03	7.774 E+03	6.6247
5	0	-4	1.47701	3.769 E+03	3.678 E+03	6.8010
3	3	0	1.47148	3.757 E+03	4.354 E+03	6.7690
4	0	10	1.46503	3.533 E+03	3.988 E+03	6.3547
2	1	-14	1.41917	3.693 E+03	3.678 E+03	13.1095
2	1	-14	1.41917	3.656 E+03	3.678 E+03	6.4814
4	2	-4	1.40069	1.750 E+03	1.384 E+03	6.1745
3	2	10	1.39046	1.761 E+03	1.820 E+03	6.1923
5	1	4	1.33509	3.096 E+03	2.866 E+03	10.6776
3	1	14	1.29192	2.810 E+03	2.410 E+03	9.5215
6	0	0	1.27434	5.689 E+03	5.291 E+03	9.5630

3	2	-14	1.19372	2.643 E+03	2.247 E+03	8.5166
4	3	10	1.10088	2.699 E+03	1.628 E+03	8.1303
5	1	-14	1.04999	2.252 E+03	1.159 E+03	6.4523

$Na_{0.3}Sr_{0.35}Zr_2P_3O_{12}$

1	0	-2	6.34223	3.221 E+04	3.044 E+04	15.7496
1	0	-2	6.34223	3.301 E+04	3.044 E+04	8.0353
1	0	4	4.56474	2.399 E+05	2.177 E+04	64.6713
1	0	4	4.56474	2.515 E+05	2.177 E+04	33.7524
1	1	0	4.40860	2.425 E+05	2.756 E+05	63.5015
1	1	3	3.81247	1.957 E+05	1.795 E+05	77.5458
1	1	3	3.81247	1.832 E+05	1.795 E+05	36.1582
2	0	-4	3.17111	3.088 E+05	2.794 E+05	45.4817
2	0	-4	3.17111	3.217 E+05	2.794 E+05	23.5989
1	1	6	2.87664	3.938 E+05	3.373 E+05	99.9999
1	1	6	2.87664	3.427 E+05	3.373 E+05	43.3569
2	1	1	2.86321	2.997 E+04	3.169 E+04	7.5574
2	1	4	2.57432	1.427 E+05	7.803 E+04	30.8429
2	1	4	2.57432	9.764 E+04	7.803 E+04	10.5179
3	0	0	2.54531	2.986 E+05	3.080 E+05	31.7630
3	0	0	2.54531	2.943 E+05	3.080 E+05	15.5987
2	0	8	2.28237	6.281 E+04	7.470 E+04	5.7613
1	1	9	2.19483	3.136 E+04	3.730 E+04	5.4685
3	0	-6	2.11408	8.557 E+04	8.273 E+04	7.1121
2	1	-8	2.02685	1.306 E+05	1.323 E+05	20.6077
2	1	-8	2.02685	1.163 E+05	1.323 E+05	9.1420
3	1	-4	1.98498	8.482 E+04	8.215 E+04	13.0431

3	1	-4	1.98498	7.044 E+04	8.215 E+04	5.3999
2	0	-10	1.95605	9.995 E+04	1.410 E+05	7.5502
2	0	-10	1.95605	2.242 E+04	1.410 E+05	8.4442
3	1	5	1.92043	6.803 E+04	6.063 E+03	10.0549
2	2	6	1.90623	2.188 E+05	2.058 E+05	32.0474
2	2	6	1.90623	1.942 E+05	2.058 E+05	14.1859
2	1	10	1.78796	2.071 E+05	2.265 E+05	28.1614
2	1	10	1.78796	1.796 E+05	2.265 E+05	12.1805
3	1	8	1.69926	1.369 E+05	8.780 E+04	17.5940
3	1	8	1.69926	1.031 E+05	8.780 E+04	6.6026
3	2	4	1.67436	1.154 E+05	1.016 E+05	14.5835
3	2	4	1.67436	1.179 E+05	1.016 E+05	7.4310
4	1	0	1.66629	2.167 E+05	1.914 E+05	27.2511
4	1	0	1.66629	2.013 E+05	1.914 E+05	12.6216
1	0	-14	1.59120	1.596 E+05	1.512 E+05	9.5461
3	1	-10	1.55096	2.127 E+05	1.580 E+05	24.7641
3	1	-10	1.55096	1.663 E+05	1.580 E+05	9.6564
4	1	6	1.52578	8.392 E+04	7.281 E+04	9.6054
4	1	-6	1.52578	8.392 E+04	7.507 E+04	9.8881
4	1	-6	1.52578	9.299 E+04	7.507 E+04	5.3083
4	1	6	1.52578	9.387 E+04	7.281 E+04	5.3586
2	0	14	1.49669	2.357 E+05	2.305 E+05	13.2243
5	0	-4	1.47505	1.081 E+05	1.063 E+05	5.9735
3	3	0	1.46953	9.580 E+04	1.149 E+04	5.2738
4	0	10	1.46307	9.654 E+04	1.200 E+05	5.2906
2	1	-14	1.41725	1.063 E+05	1.132 E+05	11.2767

2	1	-14	1.41725	9.574 E+05	1.132 E+05	5.0669
3	2	10	1.38860	4.936 E+04	6.172 E+04	5.1310
5	1	4	1.33332	9.163 E+05	9.933 E+04	9.1463
5	1	4	1.33332	1.097 E+05	9.933 E+04	5.4603
3	1	14	1.29017	7.925 E+05	8.650 E+05	7.6563
6	0	0	1.27265	1.664 E+05	1.863 E+05	7.9289
3	2	-14	1.19211	1.059 E+05	9.295 E+04	9.4650
3	2	-14	1.19211	1.385 E+05	9.295 E+04	6.1740
5	1	-14	1.04859	7.581 E+04	6.603 E+04	5.9643

$Sr_{0.5}Zr_2P_3O_{12}$

1	0	-2	6.35051	1.176 E+04	1.047 E+04	8.5015
1	0	4	4.58263	8.395 E+04	7.534 E+04	42.4390
1	0	4	4.58263	8.391 E+04	7.534 E+04	21.1602
1	1	0	4.40368	1.466 E+05	1.498 E+04	71.4325
1	1	0	4.40368	1.447 E+05	1.498 E+04	35.1789
1	1	3	3.81571	9.269 E+04	7.913 E+04	80.0710
1	1	3	3.81571	9.265 E+04	7.913 E+04	39.9416
2	0	-4	3.17525	1.384 E+04	1.072 E+04	52.5134
2	0	-4	3.17525	1.329 E+04	1.072 E+04	25.1735
1	1	6	2.88637	1.396 E+05	1.338 E+04	99.9999
1	1	6	2.88637	1.391 E+05	1.338 E+04	49.7580
2	1	1	2.86037	1.335 E+04	1.411 E+04	9.5142
1	0	-8	2.68313	4.482 E+04	1.734 E+04	15.4201
2	1	4	2.57555	3.194 E+04	2.883 E+04	21.5133
2	1	4	2.57555	3.238 E+04	2.883 E+04	10.8906
3	0	0	2.54247	1.406 E+05	1.468 E+05	47.0473

3	0	0	2.54247	1.414 E+05	1.468 E+05	23.6202
2	0	8	2.29132	2.486 E+04	2.744 E+04	7.9062
1	1	9	2.20533	1.436 E+04	1.151 E+04	8.9697
2	2	0	2.20184	2.560 E+04	2.012 E+05	7.9926
3	0	-6	2.11684	3.440 E+04	3.162 E+04	10.5499
3	0	-6	2.11684	3.314 E+04	3.162 E+04	5.0756
2	1	-8	2.03263	6.401 E+04	4.483 E+04	38.5537
2	1	-8	2.03263	4.601 E+04	4.483 E+04	13.8431
3	1	-4	1.98464	3.149 E+04	2.802 E+04	18.7682
3	1	-4	1.98464	2.750 E+04	2.802 E+04	8.1851
2	0	-10	1.96518	3.790 E+04	3.784 E+04	11.2451
2	2	6	1.90786	7.335 E+04	7.380 E+04	42.9642
2	2	6	1.90786	6.807 E+04	7.380 E+04	19.9148
4	0	-2	1.88101	2.184 E+04	1.552E+04	6.3579
2	1	10	1.79460	5.134 E+04	6.026 E+04	29.2723
2	1	10	1.79460	4.808 E+04	6.026 E+04	13.6935
3	1	-7	1.77711	9.932 E+03	1.098 E+04	5.6389
3	1	8	1.70209	2.977 E+04	2.794 E+04	16.5784
3	1	8	1.70209	2.621 E+04	2.794 E+04	7.2892
3	2	4	1.67361	3.730 E+04	3.293 E+04	20.6123
3	2	4	1.67361	3.728 E+04	3.293 E+04	10.2915
4	1	0	1.66444	6.756 E+04	6.479 E+04	37.2420
4	1	0	1.66444	6.952 E+04	6.479 E+04	19.1406
1	0	-14	1.60140	3.483 E+04	3.502 E+04	9.4295
4	0	-8	1.58763	2.861 E+04	2.703 E+04	7.7148
3	1	-10	1.55486	3.809 E+04	4.110 E+04	20.3378

3	1	-10	1.55486	3.326 E+04	4.110 E+04	8.8699
4	1	-6	1.52600	2.316 E+04	2.326 E+04	17.2576
4	1	6	1.52600	2.254 E+04	2.263 E+04	11.9275
4	1	-6	1.52600	2.068 E+04	2.326 E+04	12.2574
4	1	6	1.52600	2.013 E+04	2.263 E+04	5.3187
2	0	14	1.50498	5.160 E+04	5.106 E+04	13.5578
2	0	14	1.50498	4.916 E+04	5.106 E+04	6.4509
5	0	-4	1.47418	2.976 E+04	3.030 E+04	7.7385
4	0	10	1.46615	2.529 E+04	3.130 E+04	6.5584
3	3	0	1.46789	3.146 E+04	3.644 E+04	8.1627
2	1	-14	1.42411	2.707 E+04	2.551 E+04	13.8238
2	1	-14	1.42411	2.560 E+04	2.551 E+04	6.5293
4	2	-4	1.39793	1.380 E+04	1.242 E+04	6.9781
3	2	10	1.39108	1.459 E+04	1.460 E+04	7.3551
5	1	4	1.32241	2.715 E+04	2.502 E+04	13.3516
5	1	4	1.32241	2.715 E+04	2.502 E+04	13.3516
3	1	14	1.29509	1.743 E+04	1.711 E+04	8.4238
6	0	0	1.27123	4.107 E+04	4.351 E+04	9.8057
3	2	-14	1.19579	1.900 E+04	1.731 E+04	8.6885
4	3	10	1.10018	1.893 E+04	1.558 E+04	8.0602
5	1	-14	1.05082	1.995 E+04	9.753 E+03	8.0957

- The reflection selected from the crystallographic information framework output of the final cycle of the refinement
- Intensities less than 5% were omitted

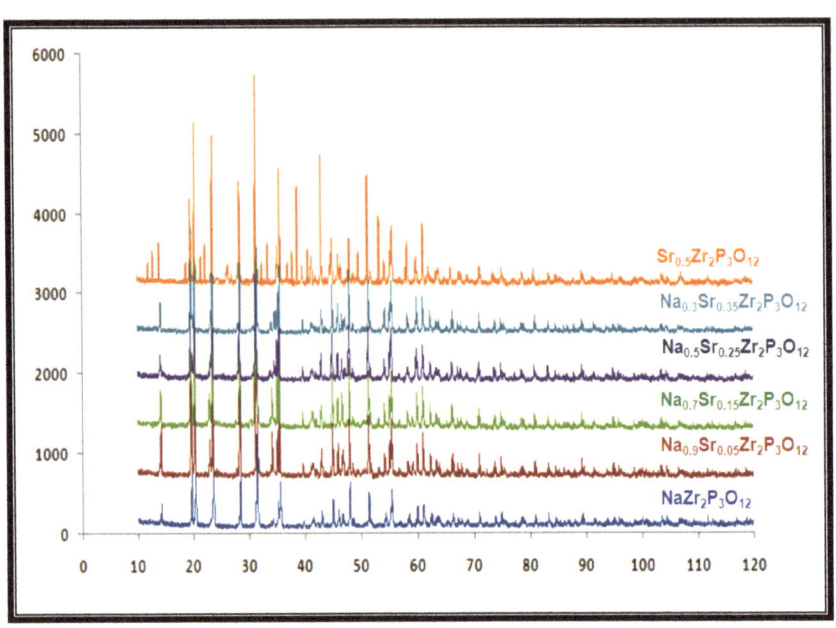

Figure 3.1 Powder XRD pattern of $Na_{1-x}Sr_{x/2}Zr_2P_3O_{12}$ (x = 0.1-1.0) ceramic samples

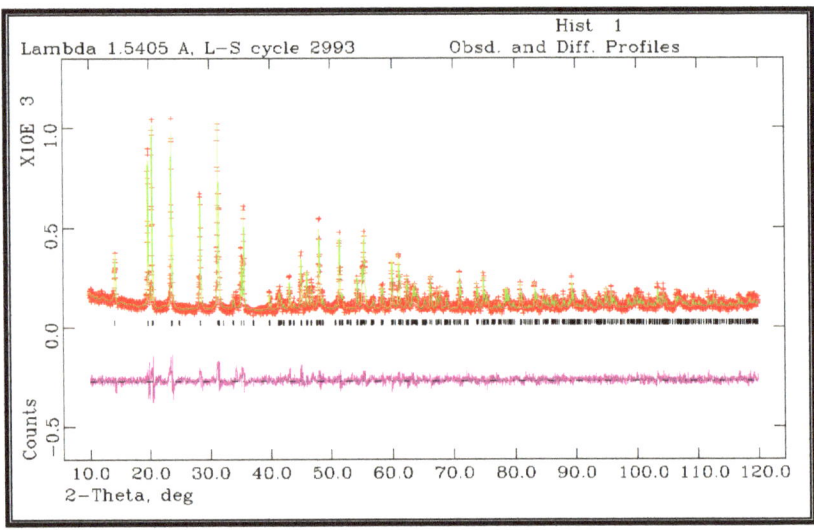

Figure 3.2 Rietveld refinement plot for $Na_{0.9}Sr_{0.05}Zr_2P_3O_{12}$ ceramic sample showing observed (+), calculated (continuous line) and difference (lower) curves. The vertical bars denote Bragg reflections of the crystalline phases

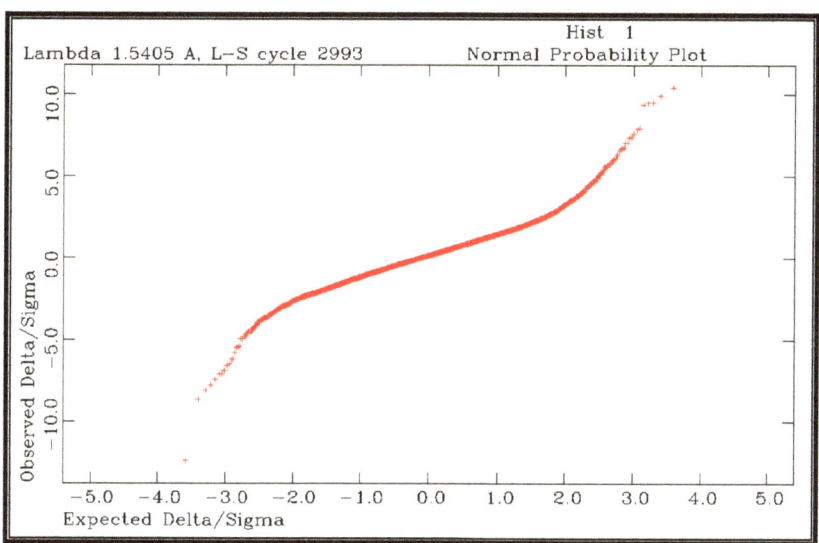

Figure 3.3 Probability plot between $I_0 - I_C$ for $Na_{0.9}Sr_{0.05}Zr_2P_3O_{12}$ ceramic sample

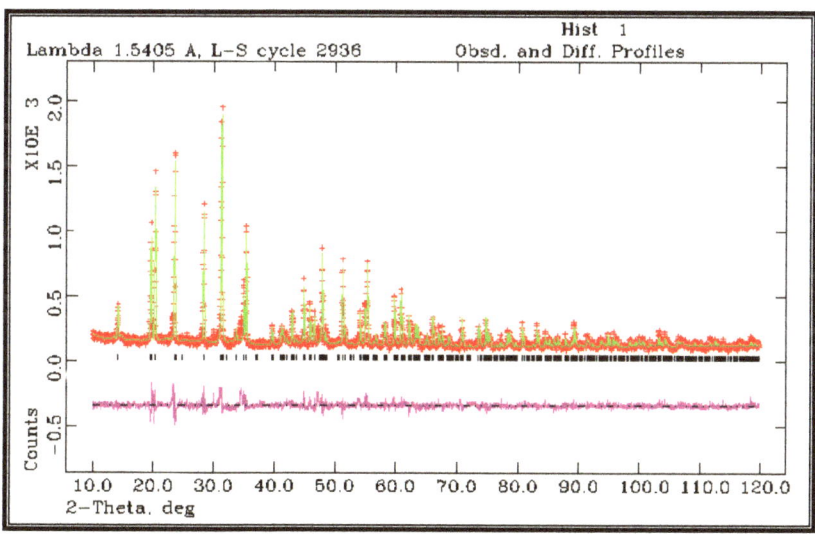

Figure 3.4 Rietveld refinement plot for $Na_{0.7}Sr_{0.15}Zr_2P_3O_{12}$ ceramic sample showing observed (+), calculated (continuous line) and difference (lower) curves. The vertical bars denote Bragg reflections of the crystalline phases

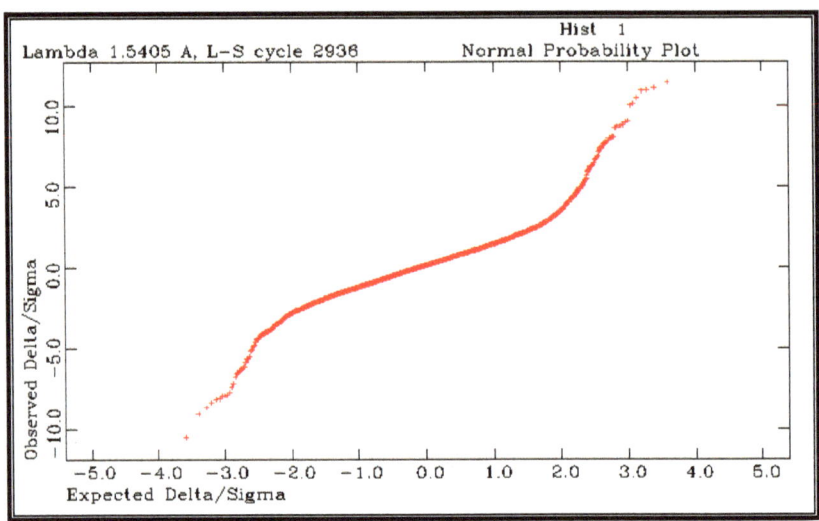

Figure 3.5 Probability plot between I_0 - I_C for $Na_{0.7}Sr_{0.15}Zr_2P_3O_{12}$ ceramic sample

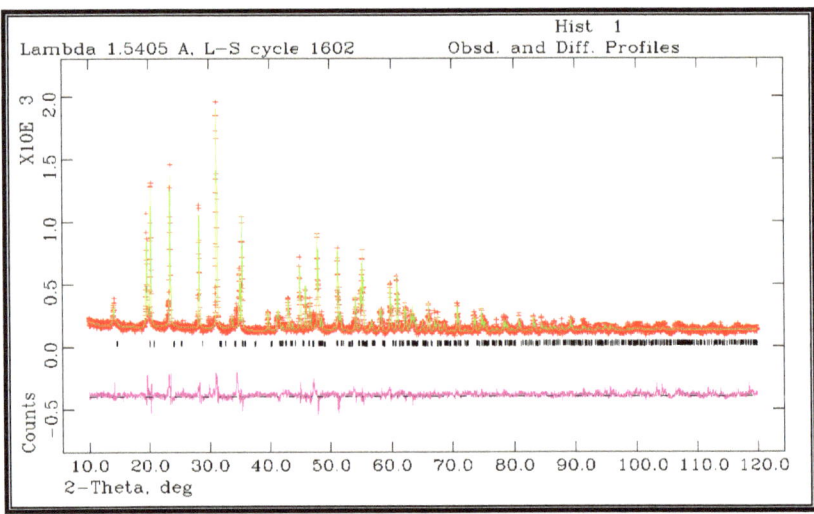

Figure 3.6 Rietveld refinement plot for $Na_{0.5}Sr_{0.25}Zr_2P_3O_{12}$ ceramic sample showing observed (+), calculated (continuous line) and difference (lower) curves. The vertical bars denote Bragg reflections of the crystalline phases

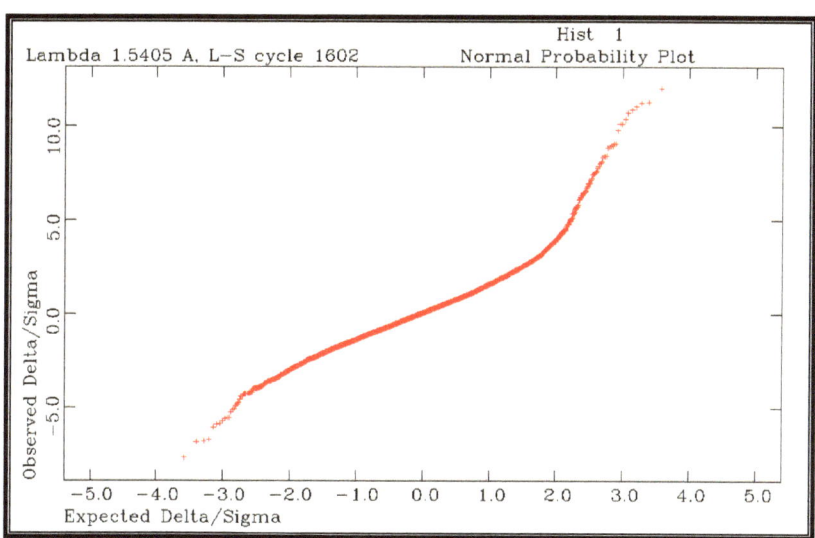

Figure 3.7 Probability plot between I_O - I_C for $Na_{0.5}Sr_{0.25}Zr_2P_3O_{12}$ ceramic sample

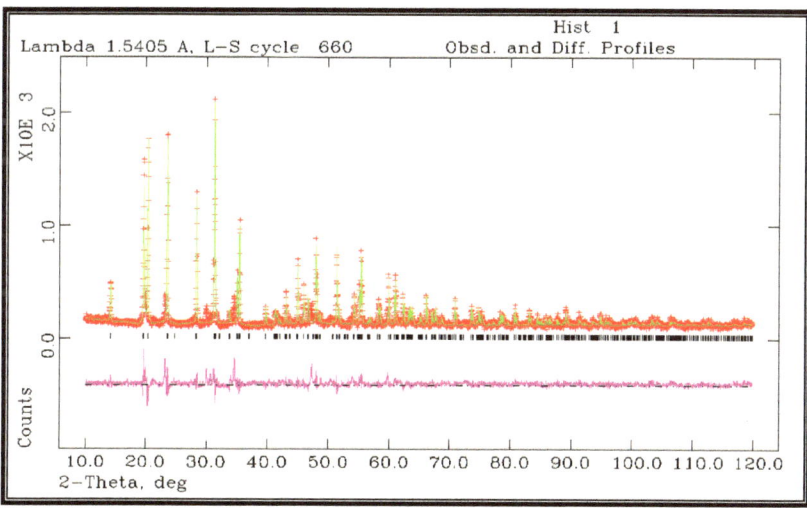

Figure 3.8 Rietveld refinement plot for $Na_{0.3}Sr_{0.35}Zr_2P_3O_{12}$ ceramic sample showing observed (+), calculated (continuous line) and difference (lower) curves. The vertical bars denote Bragg reflections of the crystalline phases

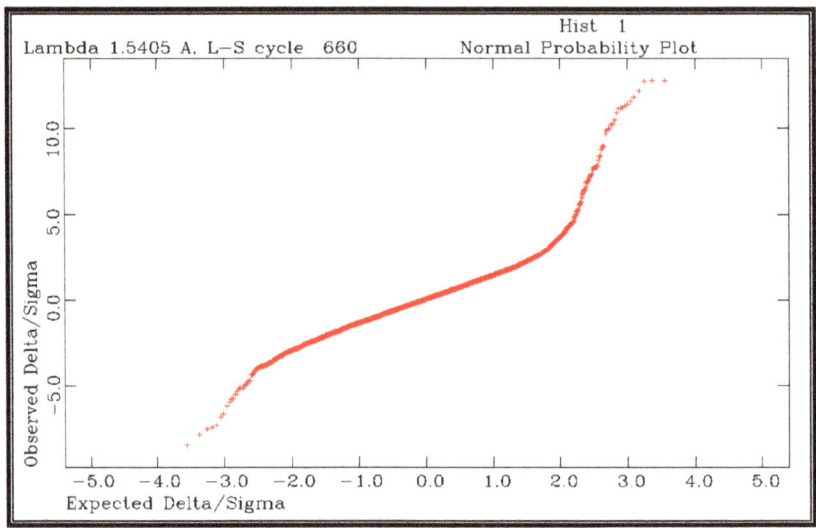

Figure 3.9 Probability plot between I_O - I_C for $Na_{0.3}Sr_{0.35}Zr_2P_3O_{12}$ ceramic sample

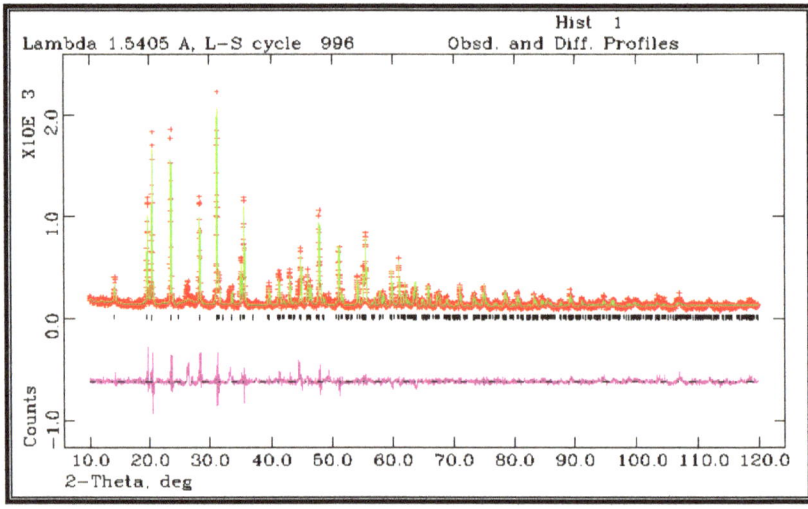

Figure 3.10 Rietveld refinement plot for $Sr_{0.5}Zr_2P_3O_{12}$ ceramic sample showing observed (+), calculated (continuous line) and difference (lower) curves. The vertical bars denote Bragg reflections of the crystalline phases

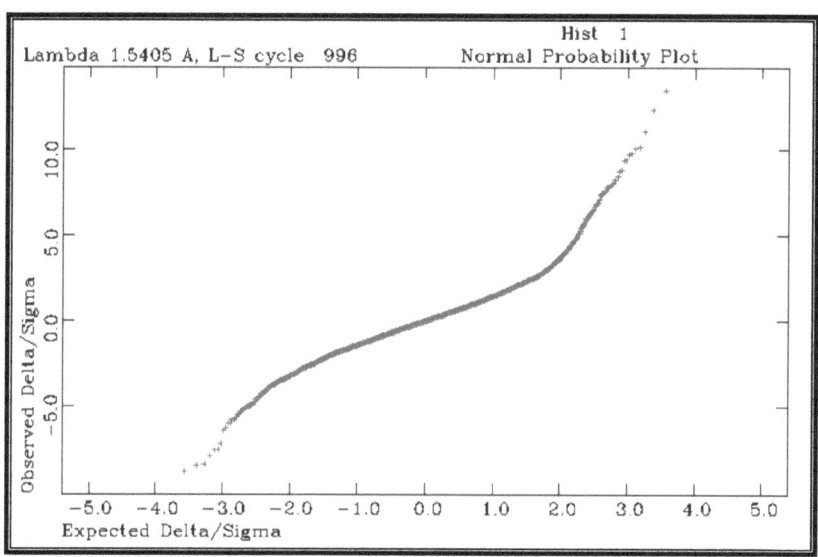

Figure 3.11 Probability plot between I_0 - I_C for $Sr_{0.5}Zr_2P_3O_{12}$ ceramic sample

Figures 3.12 (a) Scanning electron micrographs of $Na_{0.9}Sr_{0.05}Zr_2P_3O_{12}$ ceramic powder and (b) SEM of $Na_{0.9}Sr_{0.05}Zr_2P_3O_{12}$ showing location of elemental analysis

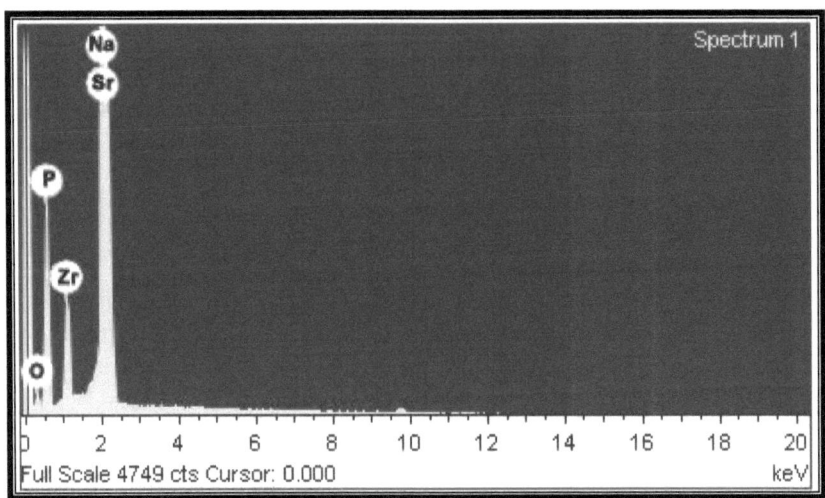

Figure 3.13 EDAX spectrum of polycrystalline mono phase of $Na_{0.9}Sr_{0.05}Zr_2P_3O_{12}$

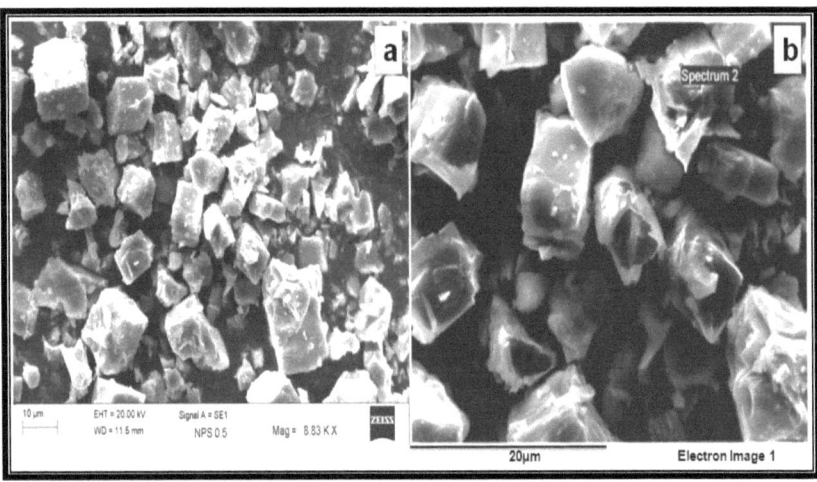

Figures 3.14 (a) Scanning electron micrographs of $Na_{0.5}Sr_{0.25}Zr_2P_3O_{12}$ ceramic powder and (b) SEM of $Na_{0.5}Sr_{0.25}Zr_2P_3O_{12}$ showing location of elemental analysis

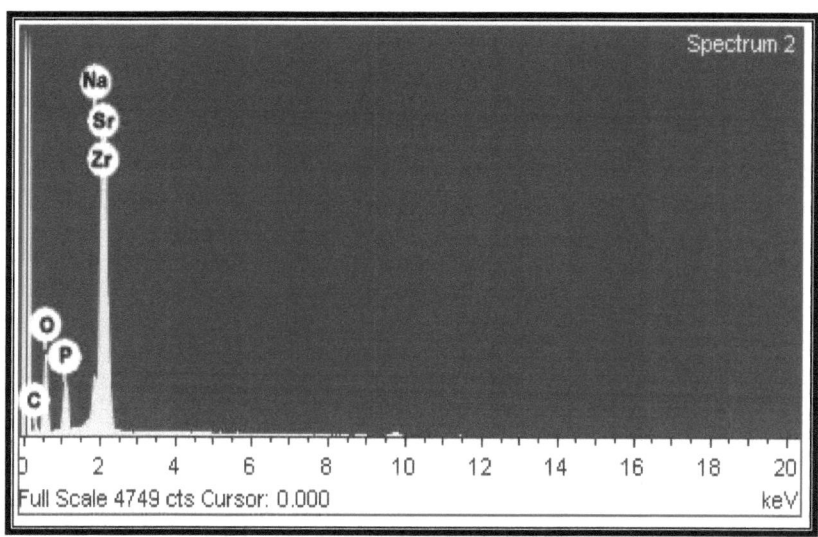

Figure 3.15 EDAX spectrum of polycrystalline mono phase of $Na_{0.5}Sr_{0.25}Zr_2P_3O_{12}$

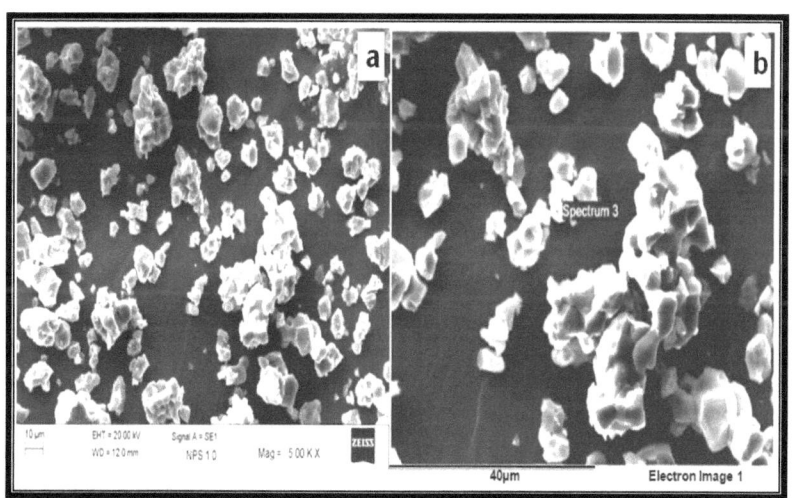

Figures 3.16 (a) Scanning electron micrographs of $Sr_{0.5}Zr_2P_3O_{12}$ ceramic powder and (b) SEM of $Sr_{0.5}Zr_2P_3O_{12}$ showing location of elemental analysis

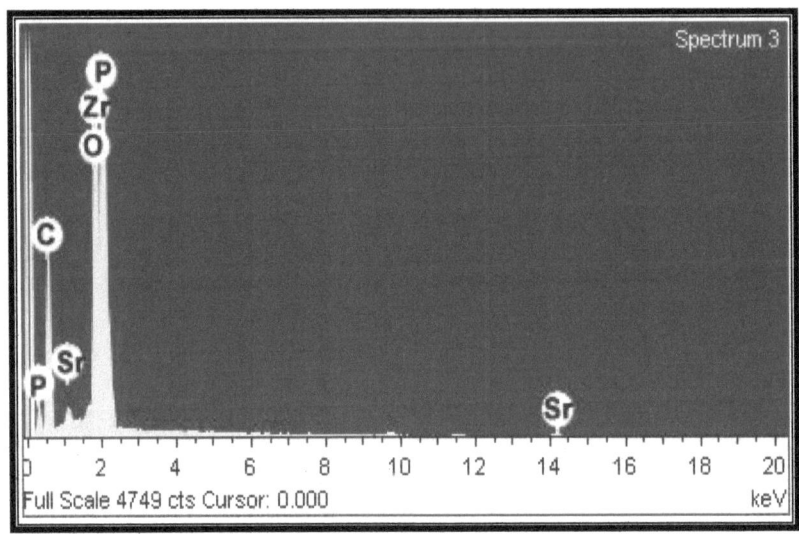

Figure 3.17 EDAX spectrum of polycrystalline mono phase of $Sr_{0.5}Zr_2P_3O_{12}$

Table 3.6 Crystallographic data for $Na_{1-x}(Cs_{1.33}Sr_1)_xZr_2P_3O_{12}$ (x = 0.1-1.0) at room temperature

Structure	Rhombohedral
Space group	R-3c
Z	6
α = β =	90°
γ =	120.0°

Parameters	$Na_{0.8}Cs_{0.266}Sr_{0.2}Zr_2P_3O_{12}$	$Na_{0.7}Cs_{0.399}Sr_{0.3}Zr_2P_3O_{12}$	$Na_{0.5}Cs_{0.665}Sr_{0.5}Zr_2P_3O_{12}$
Lattice constants			
a = b	8.81558(11)	8.82454(13)	8.80976(12)
c	22.7860(5)	22.8050(6)	22.7583(6)
Rp	0.1076	0.1213	0.1086
Rwp	0.1472	0.1681	0.1439
$R_{expected}$	0.0764	0.0747	0.0770
RF^2	0.24936	0.32010	0.21406
Volume of unit cell	1533.56(4)	1537.95(4)	1529.67(4)
S (GoF)	3.716	5.059	3.497
DWd	0.572	0.450	0.604
Unit cell formula			

weight	3226.927	3345.448	3666.221
Density$_{X-ray}$	3.494 gm/cm^3	3.612 gm/cm^3	3.980 gm/cm^3
Slope	1.5231	1.6763	1.5925

Table 3.7 Refined atomic coordinates of Na$_{1-x}$(Cs$_{1.33}$Sr$_1$)$_x$Zr$_2$P$_3$O$_{12}$ (x = 0.1-1.0) polycrystalline solid solution at room temperature

Atom	x	y	z	Occupancy	Uiso (Å^2)
Na$_{0.8}$Cs$_{0.266}$Sr$_{0.2}$Zr$_2$P$_3$O$_{12}$					
Na	0.0	0.0	0.0	0.8	0.8
Cs	0.0	0.0	0.0	0.26	0.8
Sr	0.0	0.0	0.0	0.2	0.8
Zr	0.0	0.0	0.1457	1.0	0.02007
P	0.29175	0.0	0.25	1.0	0.03151
O1	0.17594	-0.02524	0.19628	1.0	0.01093
O2	0.1925	0.1742	0.08889	1.0	0.05256
Na$_{0.7}$Cs$_{0.399}$Sr$_{0.3}$Zr$_2$P$_3$O$_{12}$					
Na	0.0	0.0	0.0	0.7	0.8
Cs	0.0	0.0	0.0	0.36	0.8
Sr	0.0	0.0	0.0	0.3	0.8
Zr	0.0	0.0	0.1457	1.0	0.01116
P	0.29175	0.0	0.25	1.0	0.02056
O1	0.17594	-0.02524	0.19628	1.0	0.00205
O2	0.1925	0.1742	0.08889	1.0	0.05996
Na$_{0.5}$Cs$_{0.665}$Sr$_{0.5}$Zr$_2$P$_3$O$_{12}$					
Na	0.0	0.0	0.0	0.5	0.8
Cs	0.0	0.0	0.0	0.665	0.8
Sr	0.0	0.0	0.0	0.5	0.8
Zr	0.0	0.0	0.14541	1.0	0.04048

P	0.29184	0.0	0.25	1.0	0.03849
O1	0.18083	-0.0232	0.19528	1.0	0.04626
O2	0.19073	0.17205	0.0875	1.0	0.05743

Table 3.8 Inter atomic distances (Å) and polyheadral distortion of polycrystalline $Na_{1-x}(Cs_{1.33}Sr_1)_xZr_2P_3O_{12}$ (x = 0.1-1.0) ceramic phase

Bond lengths(Å) / distortion (Δ)	$Na_{0.8}Cs_{0.266}Sr_{0.2}Zr_2P_3O_{12}$	$Na_{0.7}Cs_{0.399}Sr_{0.3}Zr_2P_3O_{12}$	$Na_{0.5}Cs_{0.665}Sr_{0.5}Zr_2P_3O_{12}$
Na–O2	2.59509(4)*6	2.59744(4)*6	2.55723(4)*6
Zr–O1	2.03188(2)*3	2.03383(3)*3	2.04777(3)*3
Zr–O2	2.07550(2)*3	2.07747(3)*3	2.07626(3)*3
P–O1	1.53720(2)*2	1.53858(3)*2	1.53270(3)*2
P–O2	1.52122(2)*2	1.52276(2)*2	1.53421(2)*2
ZrO_6 (Δ X 10^4)	2.30	2.30	0.9840
PO_4 (Δ X 10^4)	0.4174	0.4087	0.0037

Table 3.9 O-M-O bond angles of polycrystalline $Na_{1-x}(Cs_{1.33}Sr_1)_xZr_2P_3O_{12}$ (x = 0.1-1.0) ceramic phases

O-M-O Bond angle (Deg.)	$Na_{0.8}Cs_{0.266}Sr_{0.2}Zr_2P_3O_{12}$	$Na_{0.7}Cs_{0.399}Sr_{0.3}Zr_2P_3O_{12}$	$Na_{0.5}Cs_{0.665}Sr_{0.5}Zr_2P_3O_{12}$
O2–Na–O2	65.559(1)*6	65.568(1)*6	65.820(1)*6
O2–Na–O2	179.972(0)*2	179.980(0), 179.966(0)	180.000(0)*2
O2–Na–O2	179.980(0)	180.000(0)	180.000(0)
O2–Na–O2	114.441(1)*6	114.432(1)*6	114.180(1)*6
O1–Zr–O1	90.997 (1)*3	91.004(1)*3	92.248 (1)*3
O1–Zr–O2	92.768(1)*3	92.761(1)*3	91.253(1)*3
O1–Zr–O2	175.752 (0)*3	175.753(0)*3	174.214 (0)*3
O1–Zr–O2	90.903(1)*3	90.895(1)*3	92.216(1)*3
O2–Zr–O2	85.211(1)*3	85.219(0)*3	84.008 (1)*3
O1–P–O1	107..433(1)	107.424(2)	110.305(2)
O1–P–O2	108.595(0)*2	108.596(0)	107.280(0)*2
O1–P–O2	112.740(1)*2	112.744(1)*2	112.448(1)*2
O2–P–O2	106.829(0)	106.828(0)	107.109(0)

Table 3.10 Observed and calculated structure factors of polycrystalline Na$_{1-x}$(Cs$_{1.33}$Sr$_1$)$_x$Zr$_2$P$_3$O$_{12}$ (x = 0.1-1.0) ceramic phase. The seven columns within each group contain the values h, k, l, d-spacing, structure factor, Fosq (observed), Fcsq (calculated) and intensity respectively. The reflection selected from the CIF output of the final cycle of the refinement.

h	k	l	d-Space	F^2 (Obs.)	F^2 (Calc.)	Intensity (%)
Na$_{0.8}$Cs$_{0.266}$Sr$_{0.2}$Zr$_2$P$_3$O$_{12}$						
1	0	-2	6.34221	2.585 E+04	1.695 E+04	12.4837
1	0	-2	6.34221	2.836 E+04	1.695 E+04	6.8138
1	0	4	4.56562	2.185 E+05	2.113 E+05	55.8568
1	0	4	4.56562	2.105 E+05	2.113 E+05	26.7847
1	1	0	4.40779	2.636 E+05	2.851 E+05	62.9947
1	1	0	4.40779	2.696 E+05	2.851 E+05	32.0710
1	1	3	3.81233	1.931 E+05	1.827 E+05	70.1267
1	1	3	3.81233	1.784 E+05	1.827 E+05	32.2513
2	0	-4	3.17111	4.602 E+05	3.314 E+05	59.4063
2	0	-4	3.17111	5.385 E+05	3.314 E+05	34.5985
1	1	6	2.87708	4.620 E+05	3.045 E+05	100.000
1	1	6	2.87708	3.962 E+05	3.045 E+05	42.6921
2	1	1	2.86271	4.164 E+05	3.742 E+05	8.9325
1	0	-8	2.66858	1.470 E+05	2.438 E+04	13.9121
1	0	-8	2.66858	2.731 E+05	2.438 E+04	12.8663
2	1	4	2.57415	1.353 E+05	1.232 E+05	24.0383
2	1	4	2.57415	1.071 E+05	1.232 E+05	9.4696
3	0	0	2.54484	3.445 E+05	2.809 E+05	29.9940
3	0	0	2.54484	3.321 E+05	2.809 E+05	14.3964
2	0	8	2.28281	7.647 E+04	6.370 E+04	5.5228

1	1	9	2.19539	4.156 E+04	3.284 E+04	5.6209
1	0	10	2.18342	7.967 E+04	8.050 E+04	5.3386
3	0	-6	2.11407	1.086 E+05	6.850 E+04	6.8983
2	1	-8	2.02708	1.659 E+05	1.117 E+05	19.6686
2	1	-8	2.02708	1.453 E+05	1.117 E+05	8.5796
3	1	-4	1.98475	1.060 E+05	1.253 E+05	12.1472
3	1	-4	1.98475	1.046 E+05	1.253 E+05	5.9683
2	0	-10	1.95654	3.061 E+05	2.188 E+05	17.1341
2	0	-10	1.95654	2.901 E+05	2.188 E+05	8.0866
2	2	6	1.90617	2.504 E+05	1.850 E+05	26.8858
2	2	6	1.90617	2.173 E+05	1.850 E+05	11.6214
4	0	4	1.80975	1.198 E+05	3.170 E+04	5.9272
2	1	10	1.78828	2.496 E+05	3.115 E+05	24.2377
2	1	10	1.78828	2.207 E+05	3.115 E+05	10.6753
3	1	8	1.69930	1.133 E+05	9.099 E+04	10.1721
3	1	8	1.69930	1.198 E+05	9.099 E+04	5.3610
3	2	4	1.67413	1.503 E+05	1.506 E+05	13.2010
3	2	4	1.67413	1.697 E+05	1.506 E+05	7.4224
4	1	0	1.66599	2.714 E+05	2.113 E+05	23.6592
4	1	0	1.66599	2.521 E+05	2.113 E+05	10.9498
1	0	-14	1.59180	2.522 E+05	1.598 E+05	5.1200
4	0	-8	1.58555	1.394 E+05	6.972 E+04	5.6466
4	0	-8	1.58555	3.040 E+05	6.972 E+04	6.1357
3	1	-10	1.55110	2.238 E+05	2.345 E+05	17.5627
3	1	-10	1.55110	2.255 E+05	2.345 E+05	8.8192
4	1	6	1.52564	1.002 E+05	7.962 E+04	8.6465

4	1	-6	1.52564	1.136 E+05	8.125 E+05	8.7069
2	0	14	1.49716	2.654 E+05	2.572 E+05	9.9008
2	0	14	1.49716	2.740 E+05	2.572 E+05	5.0935
2	1	-14	1.41762	1.591 E+05	1.335 E+05	11.0073
3	2	10	1.38864	8.061 E+04	1.024 E+05	5.4224
5	1	4	1.33312	1.221 E+05	1.408 E+05	7.7839
3	1	14	1.29041	1.456 E+05	1.098 E+05	8.8990
6	0	0	1.27242	2.508 E+05	2.324 E+05	7.5316
3	2	-14	1.19227	1.532 E+05	1.258 E+05	8.5057
5	2	6	1.16369	1.092 E+05	7.241 E+04	5.8970
4	3	10	1.09936	1.505 E+05	1.522 E+05	7.6466
5	1	-14	1.04865	1.174 E+05	1.015 E+05	5.7006

$Na_{0.7}Cs_{0.399}Sr_{0.3}Zr_2P_3O_{12}$

1	0	-2	6.34830	1.747 E+04	7.686 E+03	7.8092
1	0	4	4.56972	2.129 E+05	2.007 E+05	49.3415
1	0	4	4.56972	2.107 E+05	2.007 E+05	24.2946
1	1	0	4.41227	2.665 E+05	2.922 E+05	57.5790
1	1	0	4.41227	2.663 E+05	2.922 E+05	28.6301
1	1	3	3.81603	1.938 E+05	1.805 E+05	62.6743
1	1	3	3.81603	1.737 E+05	1.805 E+05	27.9580
2	0	-4	3.17415	4.561 E+05	3.251 E+05	51.1117
2	0	-4	3.17415	5.717 E+05	3.251 E+05	31.8782
1	1	6	2.87970	5.414 E+05	3.489 E+05	100.000
1	1	6	2.87970	4.416 E+05	3.489 E+05	40.5898
2	1	1	2.86561	4.105 E+04	3.742 E+04	7.5097
1	0	-8	2.67086	1.056 E+05	3.356 E+04	8.4011

1	0	-8	2.67086	4.822 E+05	3.356 E+04	19.0872
2	1	4	2.57667	1.536 E+05	1.246 E+05	22.7652
2	1	4	2.57667	1.149 E+05	1.246 E+05	8.4740
3	0	0	2.54742	3.628 E+05	2.820 E+05	26.2866
3	0	0	2.54742	3.797 E+05	2.820 E+05	13.6896
1	1	9	2.19732	5.742 E+04	3.831 E+04	6.2223
1	0	10	2.18528	1.031 E+05	1.083 E+05	5.5257
3	0	-6	2.11610	1.313 E+05	7.015 E+04	6.6094
2	1	-8	2.02896	2.025 E+05	1.248 E+05	18.7575
2	1	-8	2.02896	2.041 E+05	1.248 E+05	9.4257
3	1	-4	1.98673	9.632 E+04	1.284 E+05	8.5815
2	0	-10	1.95826	4.078 E+05	2.671 E+05	17.6689
2	0	-10	1.95826	3.828 E+05	2.671 E+05	8.2532
2	2	6	1.90801	2.733 E+05	1.935 E+05	22.5279
2	2	6	1.90801	2.513 E+05	1.935 E+05	10.3069
4	0	4	1.81155	1.981 E+05	3.240 E+04	7.3927
2	1	10	1.78989	2.856 E+05	3.647 E+05	20.8178
2	1	10	1.78989	2.686 E+05	3.647 E+05	9.7544
3	1	8	1.70092	1.335 E+05	1.043 E+05	8.8484
3	2	4	1.67581	1.711 E+05	1.623 E+05	11.0335
3	2	4	1.67581	1.974 E+05	1.623 E+05	6.3346
4	1	0	1.66768	3.134 E+05	2.174 E+05	20.0271
4	1	0	1.66768	2.983 E+05	2.174 E+05	9.4862
3	1	-10	1.55255	3.356 E+05	2.786 E+05	9.3630
4	1	6	1.52715	1.295 E+05	9.228 E+05	7.0446
4	1	-6	1.52715	1.313 E+05	9.228 E+05	7.1412

2	0	14	1.49845	2.597 E+05	3.495 E+05	6.8292
2	1	-14	1.4886	2.056 E+05	1.822 E+05	9.8341
5	1	4	1.33446	1.262 E+05	1.551 E+05	5.4484
3	1	14	1.29157	1.683 E+05	1.435 E+05	6.8996
6	0	0	1.27371	2.889 E+05	2.553 E+05	5.7952
3	2	-14	1.19336	1.796 E+05	1.708 E+05	6.5530
4	3	10	1.10043	1.852 E+05	1.978 E+05	6.1223
5	1	-14	1.04964	2.154 E+05	1.420 E+05	6.8170

$Na_{0.5}Cs_{0.665}Sr_{0.5}Zr_2P_3O_{12}$

1	0	4	4.56098	9.631 E+04	6.442 E+04	32.3280
1	0	4	4.56098	1.136 E+04	6.442 E+04	19.0077
1	1	0	4.40488	2.264 E+05	2.487 E+05	72.4409
1	1	0	4.40488	2.302 E+05	2.487 E+05	36.7094
1	1	3	3.80928	1.258 E+05	1.105 E+05	66.7635
1	1	3	3.80928	1.200 E+05	1.105 E+05	31.7362
2	0	-4	3.16847	2.041 E+05	1.520 E+05	43.8324
2	0	-4	3.16847	2.011 E+05	1.520 E+05	21.5398
1	1	6	2.87427	2.572 E+05	2.043 E+05	100.0001
1	1	6	2.87427	2.285 E+05	2.043 E+05	44.3285
2	1	1	2.86080	2.209 E+04	2.122 E+04	8.5519
2	1	4	2.57216	9.554 E+04	5.690 E+04	33.4921
2	1	4	2.57216	7.727 E+04	5.690 E+04	13.5144
3	0	0	2.54316	1.992 E+05	1.801 E+05	34.5766
3	0	0	2.54316	2.040 E+05	1.801 E+05	17.6613
2	0	8	2.28049	3.575 E+04	4.166 E+04	5.6613
1	1	9	2.19303	1.929 E+04	2.112 E+04	5.9232

2	1	-8	2.02518	8.073 E+04	6.679 E+04	23.3384
2	1	-8	2.02518	7.528 E+04	6.679 E+04	10.8623
3	1	-4	1.98331	4.637 E+04	5.595 E+04	13.2025
3	1	-4	1.98331	4.638 E+04	5.595 E+04	6.5919
2	0	-10	1.95444	8.358 E+04	1.036 E+05	11.7735
2	0	-10	1.95444	1.519 E+04	1.036 E+05	10.6817
2	2	6	1.90464	1.359 E+05	1.012 E+05	37.6002
2	2	6	1.90464	1.205 E+05	1.012 E+05	16.6434
4	0	-2	1.88113	3.714 E+04	3.853 E+04	5.0912
2	1	10	1.78649	1.205 E+05	1.448 E+05	31.8708
2	1	10	1.78649	1.072 E+05	1.448 E+05	14.1624
3	1	8	1.69784	6.934 E+04	4.369 E+04	17.7196
3	1	8	1.69784	5.713 E+04	4.369 E+04	7.2882
3	2	4	1.67295	6.914 E+04	7.155 E+04	17.4934
3	2	4	1.67295	7.224 E+04	7.155 E+04	9.1239
4	1	0	1.66489	1.109 E+05	9.902 E+04	27.9774
4	1	0	1.66489	1.077 E+05	9.902 E+04	13.5534
1	0	-14	1.58991	7.450 E+04	7.685 E+04	9.1090
4	0	-8	1.58423	4.475 E+04	3.664 E+04	5.4580
3	1	-10	1.54968	1.040 E+05	1.021 E+05	25.0001
3	1	-10	1.54968	8.352 E+05	1.021 E+05	10.0210
4	1	-6	1.52450	4.661 E+04	3.545 E+04	11.0813
4	1	6	1.52450	4.742 E+04	3.608 E+04	11.2731
4	1	6	1.52450	4.962 E+04	3.608 E+04	5.8883
4	1	-6	1.52450	4.880 E+04	3.545 E+04	5.7908
2	0	14	1.49547	1.199 E+05	1.172 E+05	14.0733

2	0	14	1.49547	1.210 E+05	1.172 E+05	7.0884
5	0	-4	1.47381	5.905 E+04	6.717 E+04	6.8602
3	3	0	1.46829	5.246 E+04	5.310 E+04	6.0794
4	0	10	1.46185	5.711 E+04	7.608 E+04	6.5979
2	1	-14	1.41609	5.933 E+04	5.941 E+04	13.4127
2	1	-14	1.41609	5.902 E+04	5.941 E+04	6.6602
4	2	-4	1.39765	2.979 E+04	3.046 E+04	6.6732
3	2	10	1.38744	2.952 E+04	3.942 E+04	6.5788
5	1	4	1.33220	5.059 E+04	6.067 E+04	10.9536
3	1	14	1.28911	4.973 E+04	3.960 E+04	10.5108
6	0	0	1.27158	8.472 E+04	8.963 E+04	8.8614
0	0	18	1.26435	1.606 E+05	1.984 E+05	5.5767
4	2	-10	1.21797	2.514 E+04	2.594 E+04	5.0838
3	2	-14	1.19112	5.358 E+04	4.792 E+04	10.6394
3	2	-14	1.19112	6.121 E+04	4.792 E+04	6.0655
5	2	6	1.16286	3.010 E+04	2.328 E+04	5.8572
4	3	-8	1.14768	3.263 E+04	2.721 E+04	6.2745
4	3	10	1.09849	5.236 E+04	5.192 E+04	9.6654
4	2	14	1.07867	2.847 E+04	2.062 E+04	5.1601
5	1	-14	1.04772	4.428 E+04	3.298 E+04	7.7822

- The reflection selected from the crystallographic information framework output of the final cycle of the refinement
- Intensities less than 5% were omitted

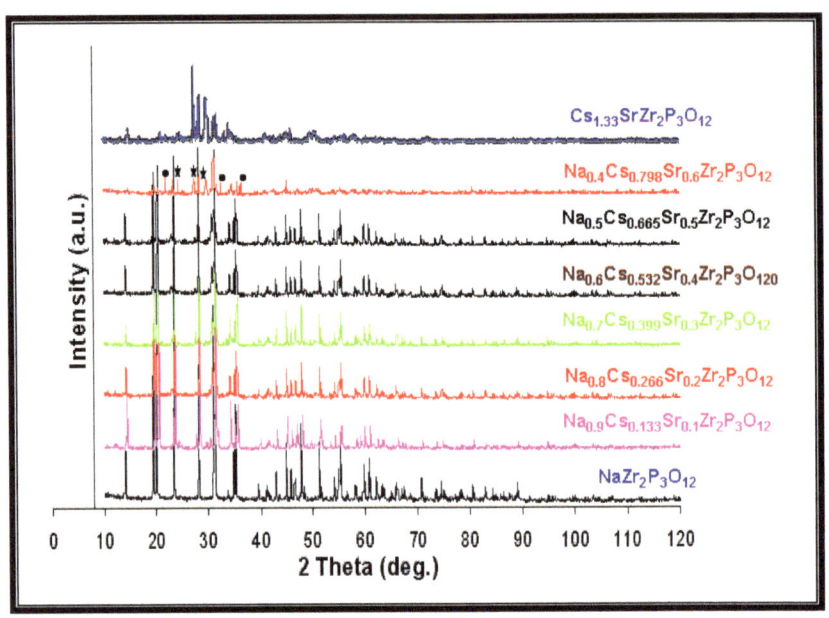

Figure 3.18 Powder XRD pattern of $Na_{1-x}(Cs_{1.33}Sr_1)_xZr_2P_3O_{12}$ (x = 0.1–1.0) ceramic samples. Marked peaks are due to $Cs_{1.33}SrZr_2P_3O_{12}$

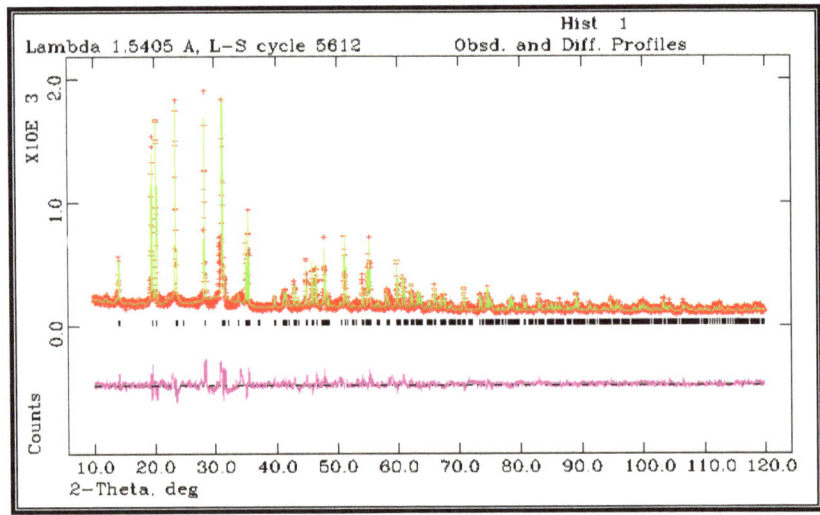

Figure 3.19 Rietveld refinement plot for $Na_{0.8}Cs_{0.266}Sr_{0.2}Zr_2P_3O_{12}$ ceramic sample showing observed (+), calculated (continuous line), and difference (lower) curves. The vertical bars denote Bragg reflections of the crystalline phases.

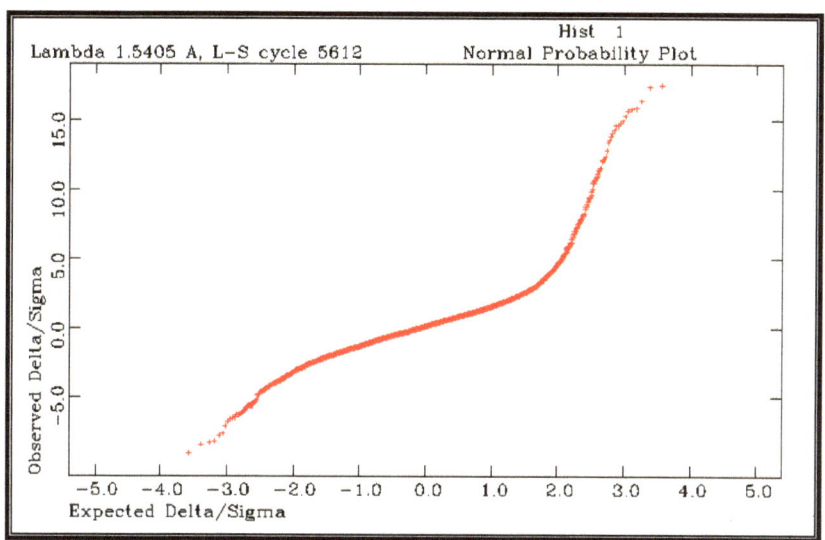

Figure 3.20 Probability plot between $I_0 - I_C$ for $Na_{0.8}Cs_{0.266}Sr_{0.2}Zr_2P_3O_{12}$ ceramic sample

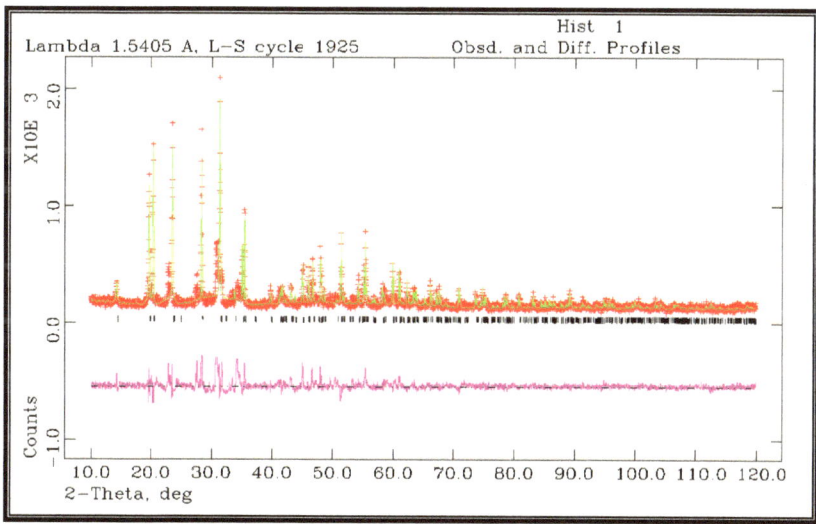

Figure 3.21 Rietveld refinement plot for $Na_{0.7}Cs_{0.399}Sr_{0.3}Zr_2P_3O_{12}$ ceramic sample showing observed (+), calculated (continuous line), and difference (lower) curves. The vertical bars denote Bragg reflections of the crystalline phases.

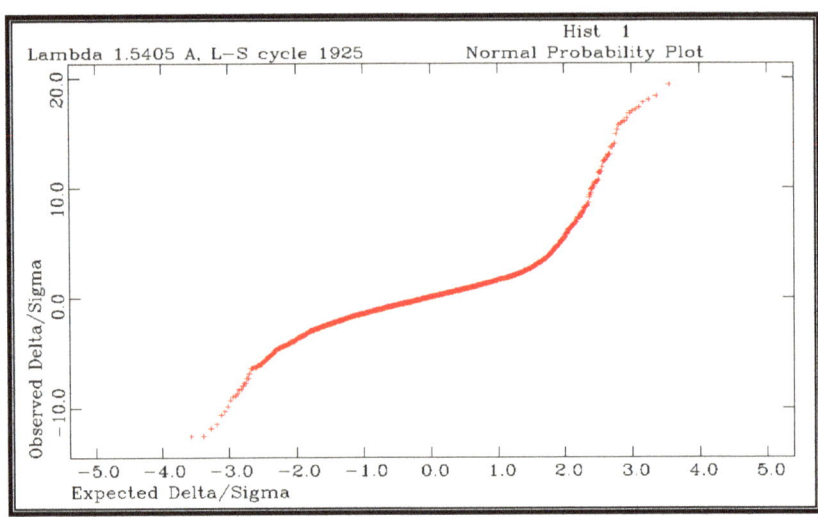

Figure 3.22 Probability plot between $I_0 - I_C$ for $Na_{0.7}Cs_{0.399}Sr_{0.3}Zr_2P_3O_{12}$ ceramic sample

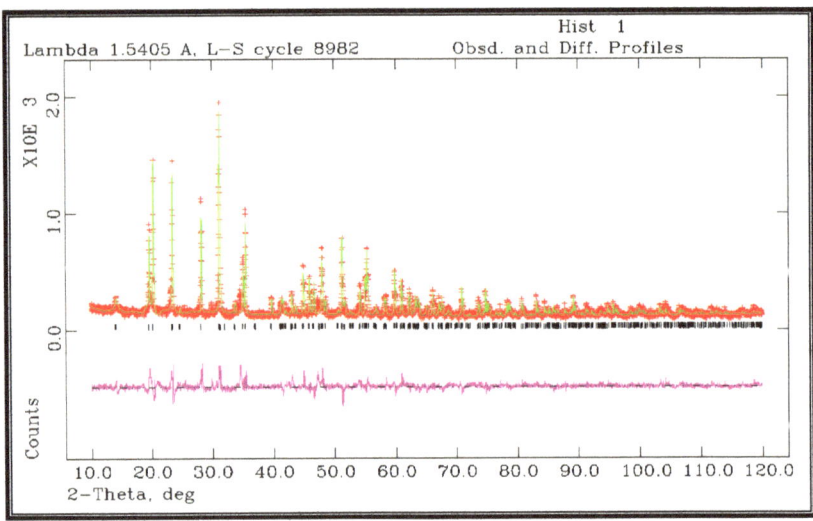

Figure 3.23 Rietveld refinement plot for $Na_{0.5}Cs_{0.665}Sr_{0.5}Zr_2P_3O_{12}$ ceramic sample showing observed (+), calculated (continuous line), and difference (lower) curves. The vertical bars denote Bragg reflections of the crystalline phases.

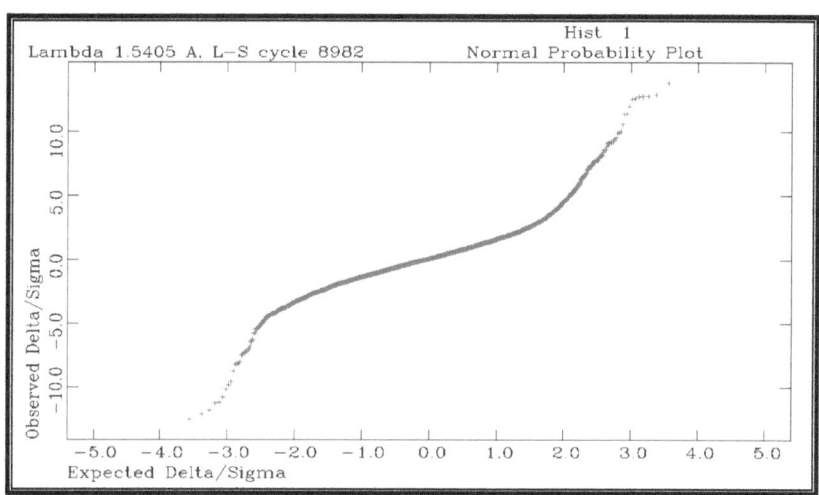

Figure 3.24 Probability plot between $I_0 - I_C$ for $Na_{0.5}Cs_{0.665}Sr_{0.5}Zr_2P_3O_{12}$ ceramic sample

Figures 3.25 (a) Scanning electron micrographs of $Na_{0.9}Cs_{0.133}Sr_{0.1}Zr_2P_3O_{12}$ ceramic powder and (b) SEM of $Na_{0.9}Cs_{0.133}Sr_{0.1}Zr_2P_3O_{12}$ showing location of elemental analysis

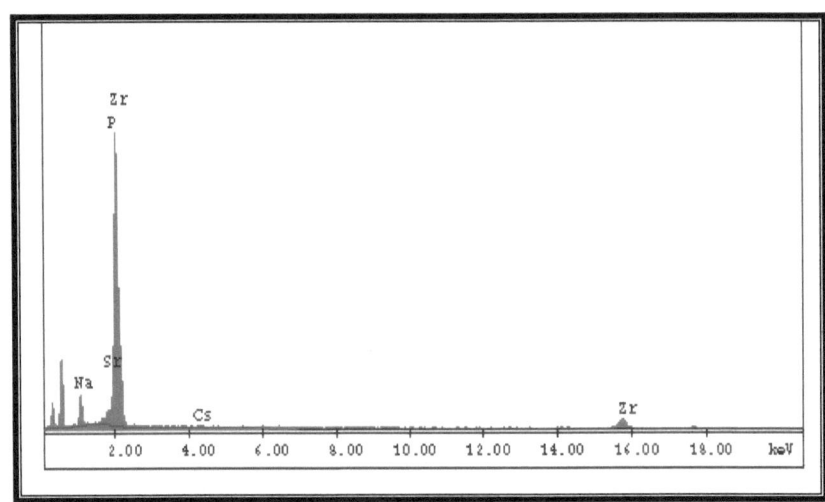

Figure 3.26 EDAX spectrum of polycrystalline mono phase of $Na_{0.9}Cs_{0.133}Sr_{0.1}Zr_2P_3O_{12}$ ceramic

Figures 3.27 (a) Scanning electron micrographs of $Na_{0.5}Cs_{0.665}Sr_{0.5}Zr_2P_3O_{12}$ ceramic powder and (b) SEM of $Na_{0.5}Cs_{0.665}Sr_{0.5}Zr_2P_3O_{12}$ showing location of elemental analysis

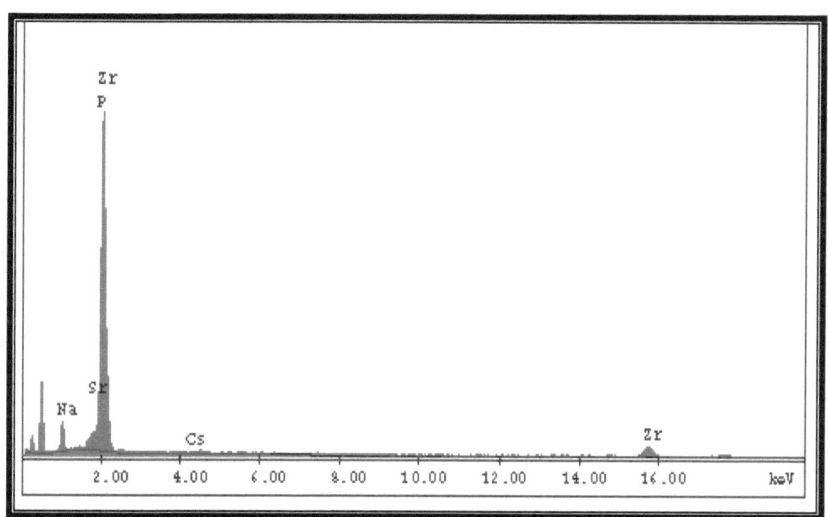

Figure 3.28 EDAX spectrum of polycrystalline mono phase of $Na_{0.5}Cs_{0.665}Sr_{0.5}Zr_2P_3O_{12}$ ceramic

Figures 3.29 (a) Scanning electron micrographs of $Cs_{1.33}Sr_1Zr_2P_3O_{12}$ ceramic powder and (b) SEM of $Cs_{1.33}Sr_1Zr_2P_3O_{12}$ showing location of elemental analysis

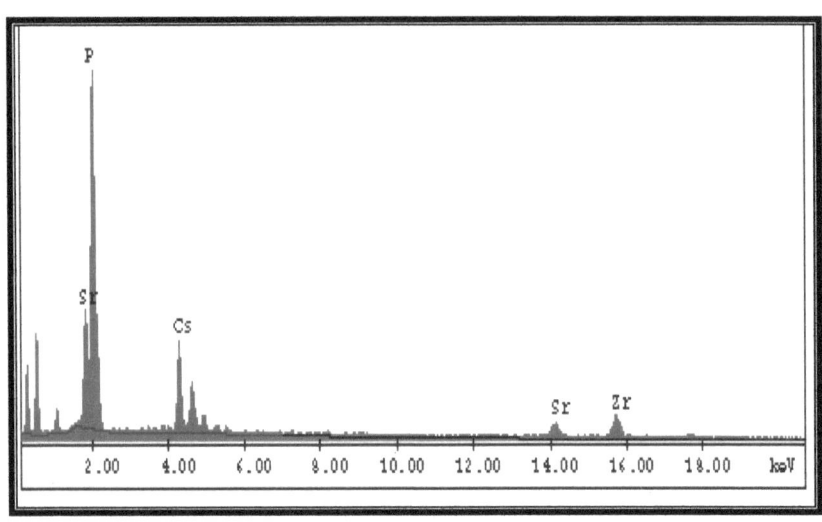

Figure 3.30 EDAX spectrum of polycrystalline mono phase of $Cs_{1.33}Sr_1Zr_2P_3O_{12}$ ceramic

Table 3.11 Crystallographic data for $Ca_{1-2x}Zr_4Mo_{2x}P_{6-2x}O_{24}$ (x = 0.1-0.5) ceramic phases

Structure	Rhombohedral
Space group	R-3
Z	6
α = β =	90°
γ =	120°

Parameters	$Ca_{0.8}Zr_4Mo_{0.2}P_{5.8}O_{24}$	$Ca_{0.4}Zr_4Mo_{0.6}P_{5.4}O_{24}$	$Zr_4MoP_5O_{24}$
Lattice constants			
a = b	8.78964(15)	8.78308(24)	8.78222(29)
c	22.6862(8)	22.7410(12)	22.7424(14)
Rp	0.0788	0.0906	0.1122
Rwp	0.1046	0.1256	0.1477
$R_{expected}$	0.0510	0.0499	0.0499
RF^2	0.08603	0.10810	0.16108
Volume of unit cell	1517.87(5)	1519.27(7)	1519.06(9)
S (GoF)	4.215	6.372	8.742
DWd	0.507	0.328	0.240
Unit cell formula weight	5886.881	5941.703	5759.502

Density$_{x-ray}$	6.440 gm/cm^3	6.494 gm/cm^3	5.897 gm/cm^3
Slope	1.7114	1.9365	1.85

Table 3.12 Refined atomic coordinates of $Ca_{1-2x}Zr_4Mo_{2x}P_{6-2x}O_{24}$ (x = 0.1-0.5) polycrystalline solid solutions at room temperature

Atom	x	y	z	Occupancy	Uiso (Å^2)
$Ca_{0.8}Zr_4Mo_{0.2}P_{5.8}O_{24}$					
Ca1	0.0	0.0	0.0	0.8	0.10917
Zr1	0.0	0.0	0.14706	1.001	0.33083
Zr2	0.0	0.0	0.64403	1.0	0.04667
Mo3	0.2928	0.0	0.2528	0.035	0.8
P4	0.2928	0.0	0.2528	0.968	0.13042
O1	0.1971	0.0059	0.1961	1.0	0.18589
O2	0.0429	-0.1673	0.6939	1.0	0.09752
O3	0.183	0.1727	0.084	1.0	0.0736
O4	0.169	-0.2097	0.5906	1.0	0.19139
$Ca_{0.4}Zr_4Mo_{0.6}P_{5.4}O_{24}$					
Ca1	0.0	0.0	0.0	0.4	0.01452
Zr1	0.0	0.0	0.14706	1.001	0.2367
Zr2	0.0	0.0	0.64403	1.0	0.04621
Mo3	0.2928	0.0	0.2528	0.035	0.00579
P4	0.2928	0.0	0.2528	0.899	0.35967
O1	0.1971	0.0059	0.1961	1.0	0.19059
O2	0.0429	-0.1673	0.6939	1.0	0.11925
O3	0.183	0.1727	0.084	1.0	0.07004
O4	0.169	-0.2097	0.5906	1.0	0.16641
$Zr_4MoP_5O_{24}$					
Zr1	0.0	0.0	0.14706	1.001	0.27315

Zr2	0.0	0.0	0.64403	1.0	0.05033
Mo3	0.2928	0.0	0.2528	0.166	0.03797
P4	0.2928	0.0	0.2528	0.832	0.64659
O1	0.1971	0.0059	0.1961	1.0	0.9131
O2	0.0429	-0.1673	0.6939	1.0	0.19776
O3	0.183	0.1727	0.084	1.0	0.08738
O4	0.169	-0.2097	0.5906	1.0	0.13692

Table 3.13 Inter atomic distances (Å) and bond valence variation of polycrystalline $Ca_{1-2x}Zr_4Mo_{2x}P_{6-2x}O_{24}$ (x = 0.1 & 0.3) ceramic phase

Bond lengths(Å)	$Ca_{0.8}Zr_4Mo_{0.2}P_{5.8}O_{24}$	$Ca_{0.4}Zr_4Mo_{0.6}P_{5.4}O_{24}$
Ca1–O3	2.46604(5)*6	2.46885(8)*6
Zr1–O1	2.03762(3)*3	2.03803(5)*3
Zr1–O3	2.12049(4)*3	2.12196(6)*3
P4/Mo3–O1	1.53199(4)	1.53426(6)
P4/Mo3–O2	1.51857(3)	1.52048(5)
P4/Mo3–O3	1.53942(3)	1.53829(4)
P4/Mo3–O4	1.51740(5)	1.51629(4)
P4/Mo3-Ca1	3.74369(6)	3.59426(9)
Ca1-Zr1	3.33626(11)*2	3.34430(17)*2

Table 3.14 Inter atomic distances (Å) and bond valence variation of polycrystalline $Zr_4MoP_5O_{24}$ ceramic phase

Bond lengths(Å)	$Zr_4MoP_5O_{24}$
Zr1–O1	2.03792(6)*3
Zr1–O3	2.12190(7)*3
P4/Mo3–O1	1.53428(7)
P4/Mo3–O2	1.52048(6)
P4/Mo3–O3	1.53814(5)
P4/Mo3–O4	1.51614(5)

Table 3.15 O-M-O bond angles of polycrystalline $Ca_{1-2x}Zr_4Mo_{2x}P_{6-2x}O_{24}$ (x=0.1 & 0.3) ceramic phases

O-M-O Bond angle (deg.)	$Ca_{0.8}Zr_4Mo_{0.2}P_{5.8}O_{24}$	$Ca_{0.4}Zr_4Mo_{0.6}P_{5.4}O_{24}$
O3–Ca1–O3	66.688(2)*6	66.546(3)*6
O3–Ca1–O3	180.0(0)*3	180.0(0)*3
O3–Ca1–O3	113.312(2)*6	113.454(3)*6
O1–Zr1–O1	93.028 (1)*3	92.915(2)*3
O1–Zr1–O3	91.099(2)*3	91.229(2)*3
O1–Zr1–O3	170.048(0)*3	170.043(0)*3
O1–Zr1–O3	95.794(1)*3	95.914(2)*3
O3–Zr1–O3	79.470(2)*3	79.334(3)*3
O2–Zr2–O2	92.070(1)*3	91.954(2)*3
O2–Zr2–O4	89.537(1)*3	89.655(2)*3
O2–Zr2–O4	88.846(1)*3	88.965(2)*3
O2–Zr2–O4	178.120(0)*3	178.118(0)*3
O4–Zr2–O4	89.520(1)*3	89.399(2)*3
O1–P4–O2	110.850(2)	111.016(3)
O1–P4–O3	107.572(1)	107.522(1)
O1–P4–O4	108.516(0)	108.495(0)
O2–P4–O3	119.055(1)	118.971(2)
O2–P4–O4	110.434(1)	110.338(2)
O3–P4–O4	109.273(0)	108.906(0)

Table 3.16 O-M-O bond angles of polycrystalline $Zr_4MoP_5O_{24}$ ceramic phases

O-M-O Bond angle(deg.)	$Zr_4MoP_5O_{24}$
O1–Zr1–O1	92.909(3)*3
O1–Zr1–O3	91.236(3)*3
O1–Zr1–O3	170.042(0)*3
O1–Zr1–O3	95.920(3)*3
O3–Zr1–O3	79.327(3)*3
O2–Zr2–O2	91.949(3)*3
O2–Zr2–O4	89.661(3)*3
O2–Zr2–O4	88.971(3)*3
O2–Zr2–O4	178.118(0)*3
O4–Zr2–O4	89.393(3)*3
O1–P4–O2	111.025(4)
O1–P4–O3	108.610(1)
O1–P4–O4	108.494(0)
O2–P4–O3	118.967(2)
O2–P4–O4	110.333(2)
O3–P4–O4	108.906(0)

Table 3.17 Selected h, k, l values, d-spacing, observed, and calculated structure factors and intensity of $Ca_{1-2x}Zr_4Mo_{2x}P_{6-2x}O_{24}$ (x = 0.1-0.5) ceramic phases

h	k	l	d-Space	F^2 (Obs.)	F^2 (Calc.)	Intensity (%)
$Ca_{0.8}Zr_4Mo_{0.2}P_{5.8}O_{24}$						
1	0	-2	6.32085	1.945 E+04	1.851 E+04	18.7278
1	0	-2	6.32085	1.931 E+04	1.851 E+04	9.2909
1	0	4	4.54809	6.901 E+04	6.749 E+04	59.7960
1	0	4	4.54809	6.669 E+04	6.749 E+04	28.8712
1	1	0	4.39488	9.350 E+04	7.761 E+04	80.3263
1	1	0	4.39488	9.277 E+04	7.761 E+04	39.8276
1	1	3	3.79980	5.777 E+04	5.658 E+04	96.1332
1	1	3	3.79980	5.603 E+04	5.658 E+04	46.5929
2	0	-4	3.16042	7.932 E+04	7.639 E+04	63.7449
2	0	-4	3.16042	7.804 E+04	7.639 E+04	31.3425
1	1	6	2.86632	6.328 E+04	5.908 E+04	99.9999
1	1	6	2.86632	6.158 E+04	5.908 E+04	48.6294
2	1	1	2.85426	9.729 E+03	9.691 E+03	15.3614
2	1	1	2.85426	8.407 E+03	9.691 E+03	6.6349
2	1	4	2.56587	1.802 E+04	1.800 E+04	27.9419
2	1	4	2.56587	1.896 E+04	1.800 E+04	14.6934
3	0	0	2.53738	5.668 E+04	6.178 E+04	43.8504
3	0	0	2.53738	5.902 E+04	6.178 E+04	22.8179
2	0	8	2.27405	9.674 E+03	1.140 E+04	7.3366
1	1	9	2.18663	5.081 E+03	5.146 E+03	7.6485
2	2	0	2.19744	7.939 E+03	6.744 E+03	5.9806
3	0	-6	2.10695	1.253 E+04	1.156 E+04	9.3637
2	1	-8	2.01969	2.050 E+04	2.084 E+04	30.3546
2	1	-8	2.01969	2.046 E+04	2.084 E+04	15.1385
3	1	-4	1.97860	1.197 E+04	1.198 E+04	17.6380
3	1	-4	1.97860	1.135 E+04	1.198 E+04	8.3593
2	0	-10	1.94876	1.447 E+04	1.504 E+04	10.6282
2	0	-10	1.94876	1.468 E+04	1.504 E+04	5.3895
2	2	6	1.89990	2.592 E+04	2.680 E+04	37.8503

2	2	6	1.89990	2.491 E+04	2.680 E+04	18.1782
4	0	-2	1.87681	7.106 E+03	6.779 E+03	5.1737
2	1	10	1.78148	2.262 E+04	2.100 E+04	32.5016
2	1	10	1.78148	2.183 E+04	2.100 E+04	15.6763
3	1	-7	1.76900	3.602 E+03	4.076 E+03	5.1666
3	1	8	1.69347	1.068 E+04	1.149 E+04	15.1286
3	1	8	1.69347	1.033 E+04	1.149 E+04	7.3096
3	2	4	1.66902	1.351 E+04	1.241 E+04	19.0574
3	2	4	1.66902	1.268 E+04	1.241 E+04	8.9364
4	1	0	1.66111	2.393 E+04	2.562 E+04	33.7124
4	1	0	1.66111	2.364 E+04	2.562 E+04	16.9398
1	0	-14	1.58498	1.273 E+04	1.304 E+04	8.8356
4	0	-8	1.58021	9.607 E+03	9.506 E+03	6.6589
3	1	-10	1.54554	1.327 E+04	1.317 E+04	18.2586
3	1	-10	1.54554	1.283 E+04	1.317 E+04	8.8171
4	1	-6	1.52082	7.335 E+03	6.465 E+03	10.0317
4	1	6	1.52082	7.475 E+03	6.589 E+03	10.2229
4	1	6	1.52082	7.357 E+03	6.589 E+03	5.0262
2	0	14	1.49098	1.642 E+04	1.691 E+04	11.1458
2	0	14	1.49098	1.665 E+04	1.691 E+04	5.6440
5	0	-4	1.47038	1.069 E+04	1.026 E+04	7.2175
3	3	0	1.46496	1.107 E+04	1.180 E+04	7.4581
4	0	10	1.45801	9.974 E+03	9.508 E+03	6.7097
2	1	-14	1.41194	8.916 E+03	8.612 E+03	11.8337
2	1	-14	1.41194	9.002 E+03	8.612 E+03	5.9673
4	2	-4	1.39441	4.686 E+03	4.545 E+03	6.1839
3	2	10	1.38384	4.464 E+03	4.423 E+03	5.8709
5	1	4	1.32911	9.824 E+03	8.553 E+03	12.6645
5	1	4	1.32911	9.140 E+03	8.553 E+03	5.8841
3	1	14	1.28548	6.561 E+03	6.390 E+03	8.3047
6	0	0	1.26869	1.585 E+04	1.325 E+04	9.9566
3	2	-14	1.18787	5.546 E+03	5.563 E+03	6.6701
4	3	10	1.09578	6.436 E+03	3.934 E+03	7.2292

$Ca_{0.4}Zr_4Mo_{0.6}P_{5.4}O_{24}$

1	0	-2	6.32220	8.041 E+04	8.553 E+04	30.4796
1	0	-2	6.32220	8.066 E+04	8.553 E+04	15.2485
1	0	4	4.55381	2.573 E+05	2.654 E+05	72.5969
1	0	4	4.55381	2.545 E+05	2.654 E+05	35.8337
1	1	0	4.39154	3.180 E+05	2.727 E+05	87.3569
1	1	0	4.39154	3.234 E+05	2.727 E+05	44.3497
1	1	3	3.79992	1.954 E+05	1.971 E+05	97.5891
1	1	3	3.79992	1.947 E+05	1.971 E+05	48.5432
2	0	-4	3.16110	3.092 E+05	3.117 E+05	89.8527
2	0	-4	3.16110	3.101 E+05	3.117 E+05	34.9777
1	1	6	2.86931	2.315 E+05	1.971 E+05	100.000
1	1	6	2.86931	2.229 E+05	1.971 E+05	48.0930
2	1	1	2.85224	3.036 E+04	3.033 E+04	13.0812
2	1	1	2.85224	2.592 E+04	3.033 E+04	5.5779
2	1	4	2.56557	9.080 E+04	8.662 E+04	37.4274
2	1	4	2.56557	8.111 E+04	8.662 E+04	35.3589
3	0	0	2.53546	2.292 E+05	2.360 E+05	47.0099
3	0	0	2.53546	2.552 E+05	2.360 E+05	26.1500
2	0	8	2.27691	3.796 E+05	3.928 E+04	7.4719
2	2	0	2.19577	3.726 E+04	3.641 E+04	7.2369
1	1	9	2.19013	1.518 E+04	1.506 E+04	5.8898
1	0	10	2.17881	2.769 E+04	2.953 E+04	5.3637
3	0	-6	2.10740	4.804 E+04	5.293 E+04	9.1921
2	1	-8	2.02137	8.237 E+04	7.459 E+04	31.0496
2	1	-8	2.02137	8.898 E+04	7.459 E+04	16.7540
3	1	-4	1.97785	5.351 E+04	5.913 E+04	20.0114
3	1	-4	1.97785	5.359 E+04	5.913 E+04	10.0122
2	0	-10	1.95179	5.509 E+04	6.582 E+04	10.2511
2	2	6	1.89996	1.085 E+05	1.060 E+05	39.9978
2	2	6	1.89996	1.158 E+05	1.060 E+05	21.3246
4	0	-2	1.87555	3.279 E+04	2.700 E+04	6.0128
2	1	10	1.78357	8.224 E+04	8.629 E+04	29.5992
2	1	10	1.78357	7.985 E+04	8.629 E+04	14.3558

3	1	8	1.69407	3.632 E+04	4.223 E+04	12.8140
3	1	8	1.69407	4.354 E+04	4.223 E+04	7.6727
3	2	4	1.66821	5.503 E+04	5.431 E+04	19.2985
4	1	0	1.65985	1.100 E+05	1.089 E+05	38.4985
4	1	0	1.65985	1.152 E+05	1.089 E+05	20.1324
1	0	-14	1.58854	5.997 E+04	5.008 E+04	10.3052
1	0	-14	1.58854	6.682 E+04	5.008 E+04	5.7355
4	0	-8	1.58055	4.893 E+04	3.987 E+04	8.3906
3	1	-10	1.54661	4.914 E+04	5.263 E+04	16.6935
3	1	-10	1.54661	4.948 E+04	5.263 E+04	16.6256
4	1	-6	1.52044	2.900 E+04	2.774 E+04	9.7786
4	1	6	1.52044	2.947 E+04	2.818 E+04	9.9358
2	0	14	1.49381	6.174 E+04	6.372 E+04	10.3251
5	0	-4	1.46957	4.198 E+04	4.528 E+04	6.9673
3	3	0	1.46385	4.849 E+04	5.456 E+04	8.0321
4	0	10	1.45879	4.662 E+04	4.204 E+04	7.7102
2	1	-14	1.41423	3.536 E+04	3.505 E+04	11.5176
2	1	-14	1.41423	3.548 E+04	3.505 E+04	5.7712
4	2	-4	1.39361	2.154 E+04	1.991 E+04	6.9631
3	2	10	1.38441	1.650 E+04	1.585 E+04	5.3151
5	1	4	1.32833	3.611 E+04	3.689 E+04	11.3729
3	1	14	1.28704	2.522 E+04	2.309 E+04	7.7962
6	0	0	1.26773	6.631 E+04	6.028 E+04	10.1509
5	2	0	1.21800	2.353 E+04	1.995 E+04	7.0133
3	2	-14	1.18897	2.582 E+04	2.635 E+04	7.5614
4	3	-8	1.14462	2.035 E+04	1.982 E+04	5.7792
4	3	10	1.09575	2.830 E+04	2.012 E+04	7.7237

$Zr_4MoP_5O_{24}$

1	0	-2	6.32188	1.150 E+05	1.231 E+05	40.3197
1	0	-2	6.32085	1.150 E+05	1.231 E+05	20.1118
1	0	4	4.55383	3.053 E+05	3.154 E+05	78.9519
1	0	4	4.55383	3.062 E+05	3.154 E+05	39.5081
1	1	0	4.39111	3.280 E+05	2.666 E+05	82.4883

1	1	0	4.39111	3.320 E+05	2.666 E+05	41.6722
1	1	3	3.79970	2.195 E+05	2.146 E+05	100.000
1	1	3	3.79970	2.200 E+05	2.146 E+05	50.0245
2	0	-4	3.16094	3.401 E+05	3.570 E+05	69.8123
2	0	-4	3.16094	3.469 E+05	3.570 E+05	35.5625
1	1	6	2.86928	2.322 E+05	1.693 E+05	90.9758
1	1	6	2.86928	2.225 E+05	1.693 E+05	43.5355
2	1	1	2.85196	3.998 E+04	3.565 E+04	53.6223
2	1	1	2.85196	3.278 E+04	3.565 E+04	6.3980
2	1	4	2.56539	9.858 E+04	1.101 E+05	36.7922
2	1	4	2.56539	9.002 E+04	1.101 E+05	16.7818
3	0	0	2.53521	2.569 E+05	2.128 E+05	47.7002
3	0	0	2.53521	2.609 E+05	2.128 E+05	24.1988
2	0	8	2.27691	3.123 E+04	2.738 E+04	5.5581
2	2	0	2.19555	2.944 E+04	2.162 E+04	5.1681
3	0	-6	2.10729	3.734 E+04	3.586 E+04	6.4540
2	2	3	2.10889	1.730 E+04	1.705 E+04	5.9811
2	1	-8	2.02133	8.531 E+04	6.612 E+04	29.0328
2	1	-8	2.02133	9.233 E+04	6.612 E+04	15.6958
3	1	-4	1.97769	5.769 E+04	6.813 E+04	19.4717
3	1	-4	1.97769	5.764 E+04	6.813 E+04	9.7184
2	0	-10	1.95183	5.951 E+04	7.195 E+04	9.9951
2	0	-10	1.95183	5.975 E+04	7.195 E+04	5.0130
2	2	6	1.89985	1.103 E+05	9.166 E+04	36.6910
2	2	6	1.89985	1.194 E+05	9.166 E+04	19.8365
4	0	-2	1.87537	5.040 E+04	4.872 E+04	8.3372
4	0	-2	1.87537	7.273 E+04	4.872 E+04	6.0107
2	1	10	1.78357	8.618 E+04	1.022 E+05	27.9738
2	1	10	1.78357	8.443 E+04	1.022 E+05	13.7708
3	1	8	1.69400	3.844 E+04	3.304 E+04	12.2245
3	1	8	1.69400	4.392 E+04	3.304 E+04	6.9767
3	2	4	1.66807	6.271 E+04	7.040 E+04	19.8208
3	2	4	1.66807	6.518 E+04	7.040 E+04	10.2907
4	1	0	1.65968	1.065 E+05	1.030 E+05	33.5849

4	1	0	1.65968	1.145 E+05	1.030 E+05	18.0449
1	0	-14	1.58863	5.712 E+05	3.657 E+05	8.8454
4	0	-8	1.58047	5.176 E+04	3.331 E+04	7.9964
3	1	-10	1.54657	4.470 E+04	6.100 E+04	14.5989
3	1	-10	1.54657	4.746 E+04	6.100 E+04	7.2560
4	1	-6	1.52033	2.856 E+04	2.243 E+04	8.6738
4	1	6	1.52033	2.856 E+04	2.243 E+04	8.7616
2	0	14	1.49387	5.881 E+04	4.920 E+04	8.8597
5	0	-4	1.46944	4.672 E+04	5.550 E+04	6.9828
4	0	10	1.45874	5.126 E+04	4.904 E+04	7.6355
3	3	0	1.46370	3.871 E+04	4.332 E+04	5.7761
2	1	-14	1.41427	3.284 E+04	2.769 E+04	9.6346
2	1	-14	1.41427	3.421 E+04	2.769 E+04	5.0259
4	2	-4	1.39349	2.267 E+04	2.846 E+04	6.5988
5	1	4	1.32821	3.665 E+04	4.427 E+04	10.3919
3	1	14	1.28705	2.332 E+04	1.521 E+04	7.1938
6	0	0	1.26760	6.667 E+04	5.792 E+04	9.1867
5	2	0	1.21787	1.885 E+04	1.390 E+04	5.0559
3	2	-14	1.18895	2.625 E+04	2.084 E+04	6.9176
4	3	-8	1.14454	1.969 E+04	1.696 E+04	5.0317
4	3	10	1.09568	3.016 E+04	2.530 E+04	7.4072

- The reflection selected from the crystallographic information framework output of the final cycle of the refinement

- Intensities less than 5% were omitted

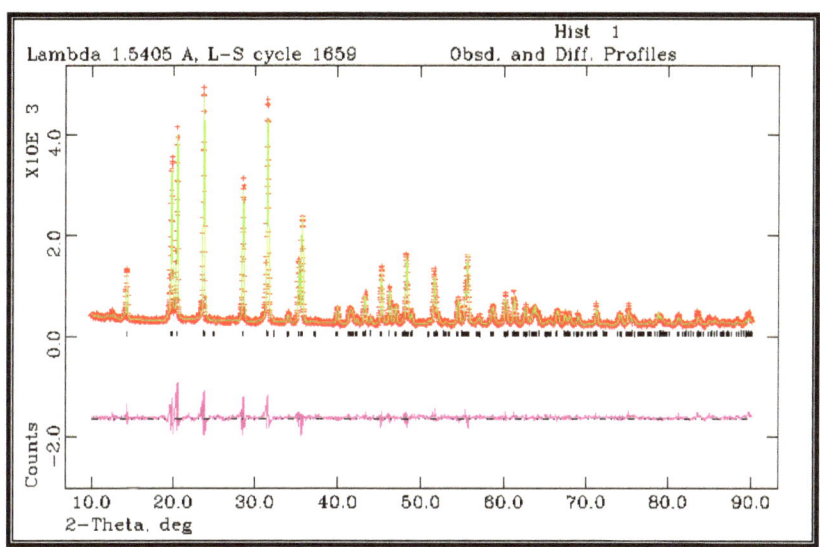

Figure 3.31 Rietveld refinement plot for $Ca_{0.8}Zr_4Mo_{0.2}P_{5.8}O_{24}$ ceramic sample showing observed (+), calculated (continuous line) and difference (lower) curves. The vertical bars denote Bragg reflections of the Crystalline phases

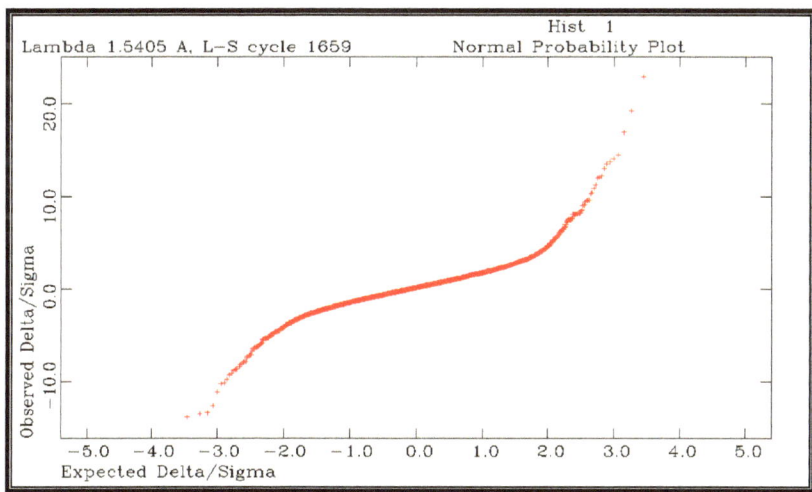

Figure 3.32 Probability plot between I_O - I_C for $Ca_{0.8}Zr_4Mo_{0.2}P_{5.8}O_{24}$ ceramic sample

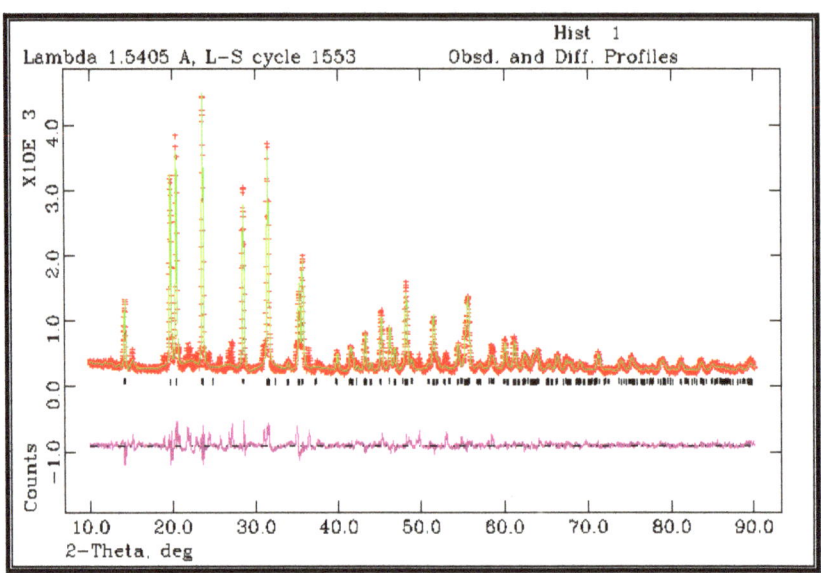

Figure 3.33 Rietveld refinement plot for $Ca_{0.4}Zr_4Mo_{0.6}P_{5.4}O_{24}$ ceramic sample showing observed (+), calculated (continuous line) and difference (lower) curves. The vertical bars denote Bragg reflections of theCrystalline phases

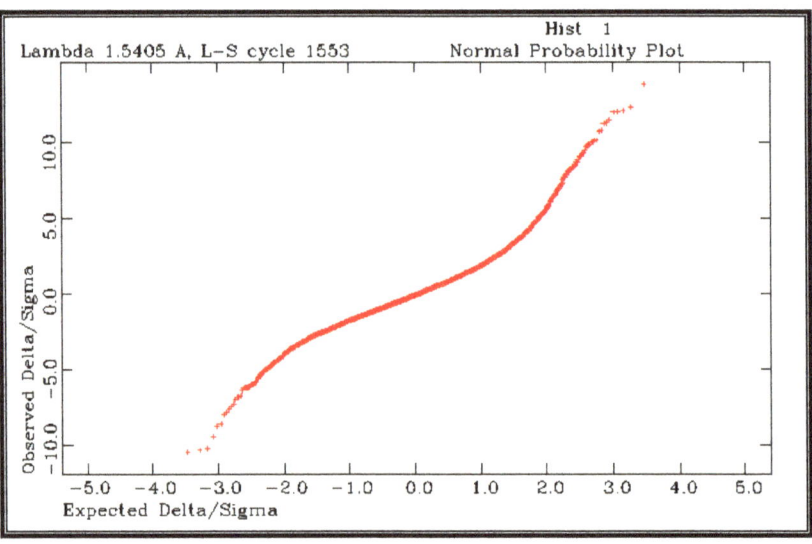

Figure 3.34 Probability plot between $I_0 - I_C$ for $Ca_{0.4}Zr_4Mo_{0.6}P_{5.4}O_{24}$ ceramic sample

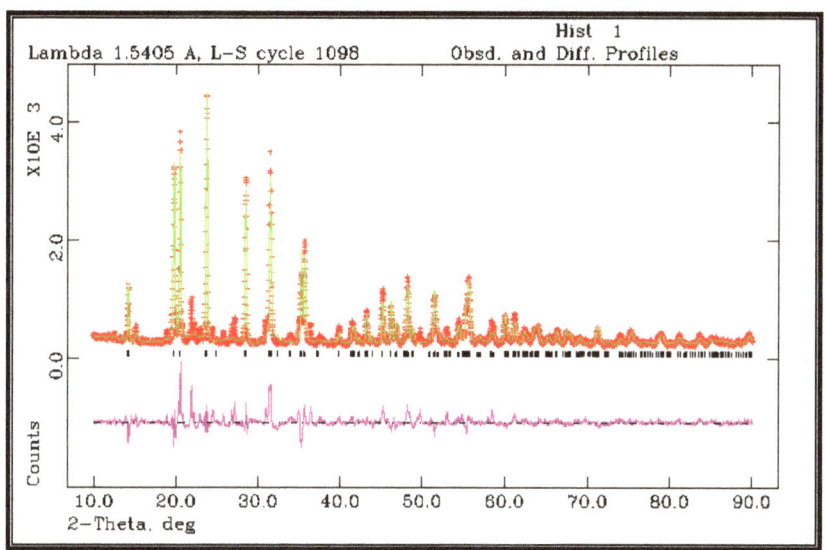

Figure 3.35 Rietveld refinement plot for $Zr_4MoP_5O_{24}$ ceramic sample showing observed (+), calculated (continuous line) and difference (lower) curves. The vertical bars denote Bragg reflections of the crystalline phases

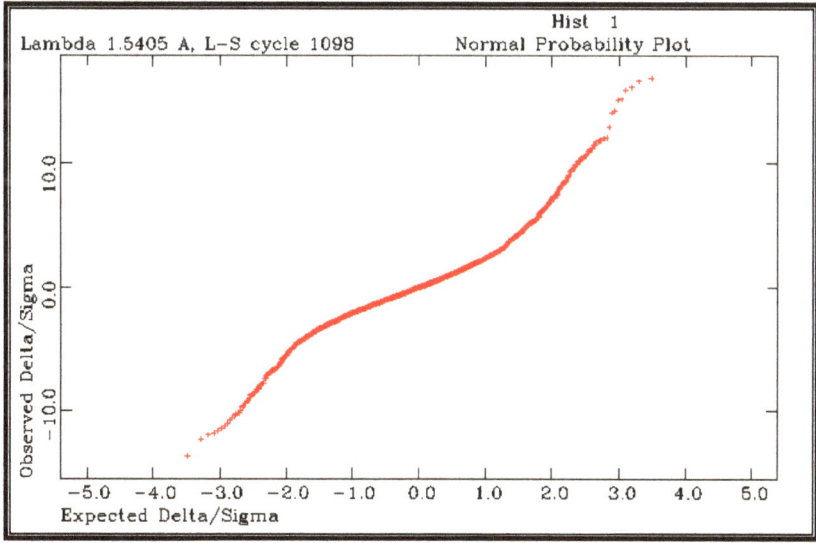

Figure 3.36 Probability plot between $I_0 - I_C$ for $Zr_4MoP_5O_{24}$ ceramic sample

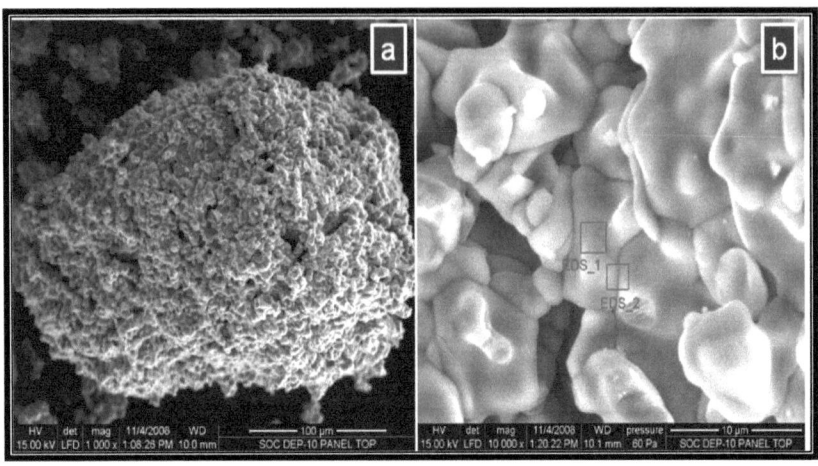

Figures 3.37 (a) Scanning electron micrographs of $Ca_{0.8}Zr_4Mo_{0.2}P_{5.8}O_{24}$ ceramic powder and (b) SEM of $Ca_{0.8}Zr_4Mo_{0.2}P_{5.8}O_{24}$ showing location of elemental analysis

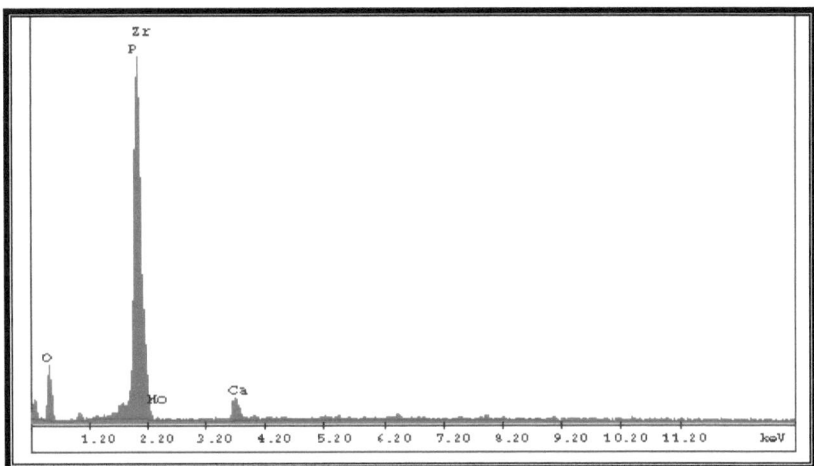

Figure 3.38 EDAX spectrum of polycrystalline mono phase $Ca_{0.8}Zr_4Mo_{0.2}P_{5.8}O_{24}$ Ceramic

Table 3.18 Crystallographic data for $Ca_{1-x}Sm_xTiO_3$ (x = 0.1 - 0.5) ceramic phases

Structure	Orthorhombic
Space group	P n m a
Z	4
$\alpha = \beta = \gamma$	90°

Parameters	$Ca_{0.9}Sm_{0.1}TiO_3$	$Ca_{0.8}Sm_{0.2}TiO_3$	$Ca_{0.7}Sm_{0.3}TiO_3$	$Ca_{0.6}Sm_{0.4}TiO_3$
Lattice constants				
a	5.43924(17)	5.4505(8)	5.4467(4)	5.4131(9)
b	7.64109(21)	7.6434(8)	7.6574(4)	7.6744(10)
c	5.38011(18)	5.3879(8)	5.39517(31)	5.4607(10)
Rp	0.1018	0.0921	0.0949	0.1156
Rwp	0.1392	0.1217	0.1231	0.1528
$R_{expected}$	0.1061	0.1065	0.0992	0.1232
RF^2	0.12792	0.19659	0.15995	0.2142
Volume of unit cell	223.607(8)	224.47(4)	225.018(14)	226.85(8)
S (GoF)	1.724	1.306	1.540	1.537
DWd	1.352	1.795	1.531	1.551
Unit cell formula weight	588.036	632.164	676.292	720.420
Density$_{X-ray}$	4.367 gm/cm^3	4.677 gm/cm^3	4.991 gm/cm^3	5.274 gm/cm^3
Slope	1.3154	1.1664	1.2319	1.2037

Table 3.19 Refined atomic coordinates of $Ca_{1-x}Sm_xTiO_3$ polycrystalline solid solution at room temperature

Atom	x	y	z	Occupancy	Uiso ($Å^2$)
$Ca_{0.9}Sm_{0.1}TiO_3$					
O1	0.317	0.527	0.326	1.0	0.09641
O2	0.018	0.25	0.037	1.0	0.02968
Ca3	0.53	0.25	0.011	0.9	0.02968
Sm4	0.53	0.25	0.011	0.1	0.05896
Ti5	0.0	0.0	0.0	1.0	0.05896
$Ca_{0.8}Sm_{0.2}TiO_3$					
O1	0.317	0.527	0.326	1.0	0.60969
O2	0.018	0.25	0.037	1.0	0.01094
Ca3	0.53	0.25	0.011	0.8	0.03007
Sm4	0.53	0.25	0.011	0.2	0.08499
Ti5	0.0	0.0	0.0	1.0	0.06083
$Ca_{0.7}Sm_{0.3}TiO_3$					
O1	0.317	0.527	0.326	1.0	0.0931
O2	0.018	0.25	0.037	1.0	0.00728
Ca3	0.53	0.25	0.011	0.7	0.03049
Sm4	0.53	0.25	0.011	0.3	0.07019
Ti5	0.0	0.0	0.0	1.0	0.04916
$Ca_{0.6}Sm_{0.4}TiO_3$					
O1	0.317	0.527	0.326	1.0	0.07104
O2	0.018	0.25	0.037	1.0	0.0051
Ca3	0.53	0.25	0.011	0.6	0.8
Sm4	0.53	0.25	0.011	0.4	0.02274
Ti5	0.0	0.0	0.0	1.0	0.03269

Table 3.20 Selected interatomic distances (Å) for $Ca_{1-x}Sm_xTiO_3$ (x=0.1-0.5) ceramic powder

$Ca_{0.9}Sm_{0.1}TiO_3$

Ca3/Sm4_O1	2.94861 (5)	Ca3_Sm4	3.83629 (10)*2
Ca3/Sm4_O1	2.73066 (5)	Ca3_Sm4	3.90942 (9)*2
Ca3/Sm4_O1	2.62362 (5)	Ca3_Sm4	3.74298 (8)*2
Ca3/Sm4_O1	2.77234 (5)	Ti5_O1	1.38192 (3)*2
Ca3/Sm4_O2	2.78840 (9)	Ti5_O1	2.46815 (6)*2
Ca3/Sm4_O2	2.65803 (8)	Ti5_O2	1.92311(5)*2
Ca3/Sm4_O2	2.94902 (10)	Ca3/Sm4_Ca3/Sm4	3.83629 (10)*2
Ca3/Sm4_O2	2.43268 (8)	Ca3/Sm4_Ca3/Sm4	3.90942 (9)*2
Ca3/Sm4_Ti5	3.45878 (8)*2	Ca3/Sm4_Ca3/Sm4	3.74298 (8)*2
Ca3/Sm4_Ti5	3.19187 (7)*2		
Ca3/Sm4_Ti5	3.35172 (8)*2		
Ca3/Sm4_Ti5	3.25534 (8)*2		

$Ca_{0.8}Sm_{0.2}TiO_3$

Ca3/Sm4_O1	2.95144 (22)	Ca3_Sm4	3.8375 (4)*2
Ca3/Sm4_O1	2.73423 (23)	Ca3_Sm4	3.9163 (4)*2
Ca3/Sm4_O1	2.95144 (22)	Ca3_Sm4	3.7497 (4)*2
Ca3/Sm4_O1	2.73423 (23)	Ti5_O1	1.38434 (14)*2
Ca3/Sm4_O2	2.7942(4)	Ti5_O1	2.47248(25)*2
Ca3/Sm4_O2	2.6636(4)	Ti5_O2	1.92373(19)*2
Ca3/Sm4_O2	2.9533(4)	Ca3/Sm4_Ca3/Sm4	3.8375(4)*2
Ca3/Sm4_O2	2.4362(4)	Ca3/Sm4_Ca3/Sm4	3.9163(4)*2
Ca3/Sm4_Ti5	3.4641(4)*2	Ca3/Sm4_Ca3/Sm4	3.7497(4)*2
Ca3/Sm4_Ti5	3.19648 (32)*2		
Ca3/Sm4_Ti5	3.25880 (33)*2		
Ca3/Sm4_Ti5	3.35536 (35)*2		

$Ca_{0.7}Sm_{0.3}TiO_3$

Ca3/Sm4_O1	2.95520 (10)	Ca3_Sm4	3.84444 (20)*2
Ca3/Sm4_O1	2.73573 (11)	Ca3_Sm4	3.91766 (17)*2

Ca3/Sm4_O1	2.62985 (9)	Ca3_Sm4	3.75064 (17)*2
Ca3/Sm4_O1	2.77777 (11)	Ti5_O1	1.38474 (6)*2
Ca3/Sm4_O2	2.79223(19)	Ti5_O1	2.47333(11)*2
Ca3/Sm4_O2	2.66168(18)	Ti5_O2	1.92722(10)*2
Ca3/Sm4_O2	2.95727(17)	Ca3/Sm4_Ca3/Sm4	3.84444(20)*2
Ca3/Sm4_O2	2.43949(14)	Ca3/Sm4_Ca3/Sm4	3.91766(17)*2
Ca3/Sm4_Ti5	3.46432(17)*2	Ca3/Sm4_Ca3/Sm4	3.75064(17)*2
Ca3/Sm4_Ti5	3.19712 (15)*2		
Ca3/Sm4_Ti5	3.36037 (14)*2		
Ca3/Sm4_Ti5	3.26370 (14)*2		

Ca$_{0.6}$Sm$_{0.4}$TiO$_3$

Ca3/Sm4_O1	2.96770 (27)	Ca3_Sm4	3.8528 (5)*2
Ca3/Sm4_O1	2.73452 (25)	Ca3_Sm4	3.9308 (5)*2
Ca3/Sm4_O1	2.96770 (27)	Ca3_Sm4	3.7602 (4)*2
Ca3/Sm4_O1	2.73452 (25)	Ti5_O1	1.38818 (16)*2
Ca3/Sm4_O2	2.7752(4)	Ti5_O1	2.48125(29)*2
Ca3/Sm4_O2	2.6454(4)	Ti5_O2	1.38818(16)*2
Ca3/Sm4_O2	2.9932(5)	Ca3/Sm4_Ca3/Sm4	3.8528 (5)*2
Ca3/Sm4_O2	2.4691(4)	Ca3/Sm4_Ca3/Sm4	3.9308 (5)*2
Ca3/Sm4_Ti5	3.3903(4)*2	Ca3/Sm4_Ca3/Sm4	3.7602 (4)*2
Ca3/Sm4_Ti5	3.2921 (4)*2		
Ca3/Sm4_Ti5	3.4519 (4)*2		
Ca3/Sm4_Ti5	3.18707 (35)*2		

*indicates the multiplicity of bond

Table 3.21 O – M – O Bond angles in $Ca_{1-x}Sm_xTiO_3$ (x = 0.1 – 0.5) ceramic powder

$Ca_{0.9}Sm_{0.1}TiO_3$

O1_Ti5_O1	180.000(0)	O1_Ca3_O2	110.577(1)*2
O1_Ti5_O2	83.405(0)*2	O1_Ca3_O2	73.717(1)*2
O1_Ti5_O2	96.595(0)*2	O1_Ca3_O2	134.373(1)*2
O2_Ti5_O2	180.000(0)	O1_Ca3_O2	123.128(1)*2
O1_Ca3_O1	64.028(2)	O1_Ca3_O2	54.623(1)*2
O1_Ca3_O1	168.787(0)	O1_Ca3_O2	72.476(1)*2
O1_Ca3_O1	111.878(2)	O2_Ca3_O2	174.108(0)
O1_Ca3_O1	91.610(2)	O2_Ca3_O2	85.587(0)
O1_Ca3_O1	122.401(2)*2	O2_Ca3_O2	88.521(0)
O1_Ca3_O1	81.003(2)		
O1_Ca3_O1	62.880(2)*2		
O1_Ca3_O1	47.775(1)		
O1_Ca3_O2	107.996(3)		
O1_Ca3_O2	135.188(1)		
O1_Ca3_O2	110.232(1)		

$Ca_{0.8}Sm_{0.2}TiO_3$

O1_Ti5_O1	179.972(0)	O1_Ca3_O2	110.597(3)*2
O1_Ti5_O2	83.417(2)*2	O1_Ca3_O2	72.469(3)*2
O1_Ti5_O2	96.583 (2)*2	O1_Ca3_O2	136.191(3)*2
O2_Ti5_O2	180.000(0)	O1_Ca3_O2	128.283(4)*2
O1_Ca3_O1	64.089(7)	O1_Ca3_O2	54.580(2)*2
O1_Ca3_O1	168.787(0)	O1_Ca3_O2	71.198(1)*2
O1_Ca3_O1	111.885(8)	O2_Ca3_O2	174.478(0)
O1_Ca3_O1	91.701(9)	O2_Ca3_O2	86.119(0)
O1_Ca3_O1	122.341(7)*2	O2_Ca3_O2	87.196(0)
O1_Ca3_O1	80.932(9)		
O1_Ca3_O1	62.913(8)*2		
O1_Ca3_O1	47.733(7)		
O1_Ca3_O2	107.156(2)		
O1_Ca3_O2	134.881(4)		
O1_Ca3_O2	109.325(4)		

Ca$_{0.7}$Sm$_{0.3}$TiO$_3$

O1_Ti5_O1	180.000(0)	O1_Ca3_O2	110.785(1)*2
O1_Ti5_O2	83.413(1)*2	O1_Ca3_O2	72.345(1)*2
O1_Ti5_O2	96.587 (1)*2	O1_Ca3_O2	135.456(1)*2
O2_Ti5_O2	180.000(0)	O1_Ca3_O2	127.265(1)*2
O1_Ca3_O1	63.985(3)	O1_Ca3_O2	54.569(1)*2
O1_Ca3_O1	168.787(0)	O1_Ca3_O2	71.598(1)*2
O1_Ca3_O1	111.837(3)	O2_Ca3_O2	175.987(0)
O1_Ca3_O1	91.585(4)	O2_Ca3_O2	86.598(0)
O1_Ca3_O1	22.437(3)*2	O2_Ca3_O2	87.856(0)
O1_Ca3_O1	80.980(4)		
O1_Ca3_O1	62.918(3)*2		
O1_Ca3_O1	47.496(2)		
O1_Ca3_O2	107.896(3)		
O1_Ca3_O2	134.396(1)		
O1_Ca3_O2	109.125(1)		

Ca$_{0.6}$Sm$_{0.4}$TiO$_3$

O1_Ti5_O1	180.000(0)	O1_Ca3_O2	110.376(3)*2
O1_Ti5_O2	83.532(2)*2	O1_Ca3_O2	74.025(4)*2
O1_Ti5_O2	96.468 (2)*2	O1_Ca3_O2	134.707(6)*2
O2_Ti5_O2	179.980(0)	O1_Ca3_O2	122.810(5)*2
O1_Ca3_O1	77.488(10)	O1_Ca3_O2	54.876(5)*2
O1_Ca3_O1	168.797(0)	O1_Ca3_O2	72.216(4)*2
O1_Ca3_O1	111.275(9)	O2_Ca3_O2	173.991(1)
O1_Ca3_O1	91.352(10)	O2_Ca3_O2	85.560(0)
O1_Ca3_O1	123.634(6)*2	O2_Ca3_O2	88.431(1)
O1_Ca3_O1	80.980(4)		
O1_Ca3_O1	62.918(3)*2		
O1_Ca3_O1	47.496(2)		
O1_Ca3_O2	107.896(3)		
O1_Ca3_O2	134.707(6)		
O1_Ca3_O2	108.112(3)		

Table 3.22 Bond distortions in CaO_8 and TiO_6 structural units of $Ca_{0.9}Sm_{0.1}TiO_3$

	CaO_8	TiO_6
Individual bond length (R_i)	2.94861	1.38192*2
	2.73066	2.46815*2
	2.62362	1.92311*2
	2.77234	
	2.78840	
	2.65803	
	2.94902	
	2.43268	
Average bond length (R_m)	2.73792	1.92439
Δ ($\times 10^{-2}$)	6.16	10.21

Table 3.23 Selected h, k, l values, d-spacing, observed, and calculated structure factors and intensity of $Ca_{1-x}Sm_xTiO_3$ (x = 0.1 – 0.5) ceramic phases

h	k	l	d-Space	F^2 (Obs.)	F^2 (Calc.)	Intensity (%)
$Ca_{0.9}Sm_{0.1}TiO_3$						
1	0	1	3.82504	4.263 E+03	4.705 E+03	4.7130
1	0	1	3.82504	4.494 E+03	4.705 E+03	2.4747
1	1	1	3.42042	1.659 E+03	3.852 E+03	3.1038
1	1	1	3.42042	1.550 E+03	3.852 E+03	1.4446
2	0	0	2.71962	9.312 E+04	9.611 E+04	31.8299
2	0	0	2.71962	9.680 E+04	9.611 E+04	16.4924
1	2	1	2.70312	7.371 E+04	7.517 E+04	100.000
1	2	1	2.70312	7.263 E+04	7.517 E+04	49.1121
0	0	2	2.69005	7.006 E+04	7.443 E+04	23.6161
0	0	2	2.69005	6.697 E+04	7.443 E+04	11.2350
0	0	2	2.69005	7.006 E+04	7.443 E+04	23.6161

2	1	0	2.56217	5.077 E+03	7.605 E+03	3.2213
2	1	0	2.56217	4.956 E+03	7.605 E+03	1.5676
1	0	2	2.41128	2.323 E+03	3.198 E+03	1.3699
2	1	1	2.31325	1.156 E+03	8.358 E+02	1.2992
0	3	1	2.30209	1.276 E+04	3.086 E+03	3.5563
2	2	0	2.21561	1.422 E+04	1.218 E+04	7.6105
2	2	0	2.21561	1.580 E+04	1.218 E+04	4.2185
0	2	2	2.19953	1.193 E+04	9.385 E+04	6.3381
0	2	2	2.19953	1.164 E+04	9.385 E+04	3.0817
1	3	1	2.12002	2.638 E+03	1.334 E+03	2.6915
2	2	1	2.04869	2.192 E+03	2.174 E+03	2.1562
2	2	1	2.04869	2.318 E+03	2.174 E+03	1.1372
1	2	2	2.03912	1.486 E+03	1.666 E+03	1.4553
1	2	2	2.03912	2.672 E+03	1.666 E+03	1.3046
0	4	0	1.91027	1.245 E+05	1.156 E+05	28.5008
2	0	2	1.91252	8.509 E+04	7.559 E+04	38.9972
0	4	0	1.91027	1.182 E+05	1.156 E+05	13.4976
2	0	2	1.91252	7.866 E+04	7.559E+04	17.9807
2	3	0	1.85904	4.222 E+03	1.528 E+03	1.8812
2	1	2	1.85529	3.948 E+03	2.431 E+03	3.5109
2	1	2	1.85529	2.980 E+03	2.431 E+03	1.3218
0	1	3	1.74593	3.547 E+03	1.945 E+03	1.4870
2	2	2	1.71021	1.617 E+03	9.093 E+02	1.3295
3	1	1	1.67629	4.926 E+03	2.946 E+03	3.9755
3	1	1	1.67629	5.305 E+03	2.946 E+03	2.1359
3	2	1	1.56698	2.094 E+04	1.827 E+04	15.8886

2	4	0	1.56319	4.197 E+04	4.149 E+04	15.8870
3	2	1	1.56698	1.844 E+04	1.827 E+04	6.9790
2	4	0	1.56319	4.049 E+04	4.149 E+04	7.6469
0	4	2	1.55751	3.869 E+04	3.891 E+04	14.5985
0	4	2	1.55751	4.669 E+04	3.891 E+04	8.7892
1	2	3	1.55561	2.129 E+04	1.868 E+04	16.0496
1	2	3	1.55561	2.495 E+04	1.868 E+04	9.3828
1	2	3	1.52937	1.607 E+03	2.603 E+03	1.1930
2	1	3	1.46923	1.510 E+03	1.146 E+03	1.0816
3	3	1	1.42437	2.781 E+03	1.816 E+03	1.9378
3	3	1	1.42437	3.027 E+03	1.816 E+03	1.0524
4	0	0	1.35981	3.084 E+04	2.466 E+04	5.1584
4	0	0	1.35981	3.109 E+04	2.466 E+04	2.5947
2	4	2	1.35156	3.005 E+04	3.054 E+04	19.9993
2	4	2	1.35156	3.025 E+04	3.054 E+04	10.0436
0	0	4	1.34503	2.100 E+04	1.898 E+04	3.4779
4	1	0	1.33878	4.923 E+03	4.975 E+03	1.6242
0	0	4	1.34503	2.220 E+04	1.898 E+04	1.8346
3	1	3	1.25763	1.681 E+03	1.197 E+03	1.0494
4	0	2	1.21356	1.610 E+04	1.538 E+04	4.8666
4	0	2	1.21356	1.439 E+04	1.538 E+04	2.1690
3	2	3	1.20944	1.123 E+04	1.192 E+04	6.7694
1	6	1	1.20830	1.160 E+04	1.233 E+04	6.9863
3	2	3	1.20944	1.011 E+04	1.192 E+04	3.0397
1	6	1	1.20830	1.075 E+04	1.233 E+04	3.2302
2	0	4	1.20564	1.249 E+04	1.461 E+04	3.7537

4	3	0	1.19956	3.903 E+03	4.992 E+03	1.1671
2	0	4	1.20564	1.259 E+04	1.461 E+04	1.8875
4	1	2	1.19855	2.684 E+03	3.660 E+03	1.6038
2	1	4	1.19091	2.042 E+03	6.658 E+02	1.2132
2	1	4	1.19091	3.408 E+03	6.658 E+02	1.0100
3	5	1	1.14188	2.397 E+03	1.145 E+03	1.3689
3	3	3	1.14014	2.390 E+03	1.066 E+03	1.3629
4	4	0	1.10780	1.333 E+04	1.144 E+04	3.6963
4	4	0	1.10780	1.092 E+04	1.144 E+04	1.5101
0	4	4	1.09977	1.246 E+04	1.028 E+04	3.4301
4	3	2	1.09557	2.951 E+03	2.266 E+03	1.6188
0	4	4	1.09977	1.338 E+04	1.028 E+04	1.8380

$Ca_{0.8}Sm_{0.2}TiO_3$

0	1	1	4.40379	6.976 E+01	1.684 E+01	1.1847
1	0	1	3.83179	1.677 E+02	5.843 E+02	2.6187
1	0	1	3.83179	8.965 E+01	5.843 E+02	1.3432
1	1	1	3.42545	9.411 E+01	3.870 E+01	1.3854
2	0	0	2.72527	8.300 E+03	7.543 E+03	55.4766
2	0	0	2.72527	7.904 E+03	7.543 E+03	26.3903
1	2	1	2.70592	3.751 E+03	3.792 E+03	100.00
1	2	1	2.70592	3.872 E+03	3.792 E+03	51.5743
0	0	2	2.69397	4.435 E+03	4.527 E+03	29.5101
0	0	2	2.69397	3.961 E+03	4.527 E+03	13.1661
2	1	0	2.56698	7.572 E+02	3.893 E+02	9.8996
2	1	0	2.56698	7.393 E+02	3.893 E+02	4.8279
2	0	1	2.43188	1.073 E+02	3.399 E+01	1.3759

1	0	2	2.41508	9.277 E+01	5.595 E+01	1.1868
2	1	1	2.31741	1.098 E+02	2.900 E+01	2.7685
2	1	1	2.31741	1.119 E+02	2.900 E+01	1.4103
0	3	1	2.30238	2.729 E+02	4.170 E+01	3.4337
0	3	1	2.30238	3.070 E+02	4.170 E+01	1.9299
2	2	0	2.21889	7.740 E+02	4.126 E+02	9.6164
2	2	0	2.21889	9.998 E+02	4.126 E+02	6.2059
0	2	2	2.20190	6.818 E+02	3.112 E+02	8.4489
0	2	2	2.20190	8.072 E+02	3.112 E+02	4.9974
1	3	1	2.12162	5.772 E+01	3.482 E+01	1.4127
2	2	1	2.05171	1.996 E+02	5.167 E+02	4.8303
2	2	1	2.05171	1.766 E+02	5.167 E+02	2.1354
2	0	2	1.91590	4.638 E+03	3.312 E+03	54.8282
2	0	2	1.91590	3.742 E+03	3.312 E+03	22.0977
0	4	0	1.91086	5.798 E+03	5.063 E+03	34.2366
0	4	0	1.91086	4.882 E+03	5.063 E+03	14.4020
2	3	0	1.86115	2.276 E+02	2.086 E+02	1.3308
2	1	2	1.85840	1.516 E+02	1.527 E+02	3.5464
2	1	2	1.85840	2.543 E+02	1.527 E+02	2.9712
2	3	1	1.75916	1.097 E+02	1.566 E+01	2.5164
0	1	3	1.74837	1.677 E+02	9.546 E+01	1.9194
2	2	2	1.71273	7.857 E+01	1.580 E+02	1.7845
2	2	2	1.71273	1.202 E+02	1.580 E+02	1.3639
3	1	1	1.67953	3.744 E+02	1.592 E+02	8.4420
3	1	1	1.67953	3.974 E+02	1.592 E+02	4.4760
3	2	1	1.56968	1.213 E+03	1.133 E+03	26.6236

3	2	1	1.56968	1.094 E+03	1.133 E+03	11.9932
2	4	0	1.56458	2.639 E+03	2.800 E+03	28.9264
2	4	0	1.56458	2.438 E+03	2.800 E+03	13.3488
0	4	2	1.55859	2.084 E+03	2.337 E+03	22.7993
1	2	3	1.55765	9.216 E+02	9.773 E+03	20.1643
0	4	2	1.55859	2.488 E+03	2.337 E+03	13.6002
1	2	3	1.55765	1.114 E+02	9.773 E+03	12.1734
2	3	2	1.53126	2.202 E+02	1.053 E+02	4.7824
2	3	2	1.53126	1.817 E+02	1.053 E+02	1.9716
2	0	3	1.49963	1.687 E+02	6.532 E+01	1.8154
2	0	3	1.49963	2.011 E+02	6.532 E+01	1.0808
2	1	3	1.47157	1.550 E+02	6.447 E+01	3.3074
0	3	3	1.46753	2.189 E+02	7.071 E+01	1.1654
3	3	1	1.42647	2.561 E+02	1.015 E+02	5.3854
3	3	1	1.42647	2.136 E+02	1.015 E+02	2.2428
2	2	3	1.39600	7.234 E+01	5.543 E+01	1.5048
2	2	3	1.39600	1.685 E+01	5.543 E+01	1.7504
4	0	0	1.36264	1.364 E+03	1.422 E+03	7.0029
4	0	0	1.36264	1.924 E+03	1.422 E+03	4.9331
2	4	2	1.35296	1.681 E+03	1.454 E+03	34.4033
2	4	2	1.35296	1.616 E+03	1.454 E+03	16.5178
0	0	4	1.34699	9.795 E+02	9.206 E+02	4.9991
0	0	4	1.34699	9.260 E+02	9.206 E+02	2.3600
4	1	0	1.34148	3.108 E+02	3.186 E+02	3.1652
4	1	0	1.34148	3.452 E+02	3.186 E+02	1.7556
2	5	0	1.33326	1.306 E+02	8.913 E+01	1.3259

4	0	1	1.32104	3.398 E+02	2.787 E+01	1.7129
4	1	1	1.30174	6.500 E+01	5.709 E+01	1.3015
2	3	3	1.29238	1.501 E+02	4.784 E+01	2.9918
2	3	3	1.29238	1.799 E+02	4.784 E+01	1.7907
0	2	4	1.27039	2.828 E+02	5.801 E+01	1.3923
3	1	3	1.25980	1.615 E+02	5.429 E+01	3.1695
3	1	3	1.25980	1.369 E+02	5.429 E+01	1.3407
4	2	1	1.24855	5.317 E+01	2.282 E+01	1.0372
4	2	1	1.24855	2.155 E+01	2.282 E+01	2.0983
1	2	4	1.23723	2.918 E+01	1.063 E+01	1.1603
4	0	2	1.21594	9.561 E+02	7.831 E+02	9.1587
4	0	2	1.21594	1.033 E+02	7.831 E+02	4.9391
3	2	3	1.21140	5.091 E+02	3.761 E+02	9.7281
1	6	1	1.20885	6.847 E+02	5.312 E+02	13.0631
3	2	3	1.21140	4.697 E+02	3.761 E+02	4.4799
2	0	4	1.20754	5.961 E+02	6.172 E+02	2.8360
1	6	1	1.20885	5.675 E+02	5.312 E+02	5.4339
2	0	4	1.20754	7.340 E+02	6.172 E+02	6.9968
4	3	0	1.20158	1.910 E+02	2.186 E+02	1.8144
4	1	2	1.20084	1.663 E+02	1.821 E+02	3.1573
4	3	0	1.20158	2.227 E+02	2.186 E+02	1.0557
4	1	2	1.20084	2.054 E+02	1.821 E+02	1.9463
2	5	2	1.19493	1.031 E+02	5.099 E+01	1.9507
2	1	4	1.19275	8.598 E+01	3.137 E+01	1.6243
2	5	2	1.19493	1.620 E+02	5.099 E+01	1.5292
2	1	4	1.19275	1.290 E+01	3.137 E+01	1.2162

4	3	1	1.17277	7.819 E+01	4.128 E+01	1.4582
0	5	3	1.16409	3.088 E+01	3.958 E+01	2.8627
3	5	1	1.14309	1.508 E+02	4.515 E+01	2.7533
3	3	3	1.14182	1.335 E+02	3.851 E+01	2.4358
3	5	1	1.14309	1.536 E+02	4.515 E+01	1.3992
3	3	3	1.14182	1.543 E+02	3.851 E+01	1.4050
2	6	1	1.12845	5.883 E+01	4.545 E+01	1.0624
4	4	0	1.10944	7.453 E+02	6.378 E+02	6.6266
4	4	0	1.10944	9.218 E+02	6.378 E+02	4.0887
0	4	4	1.10095	9.399 E+02	4.844 E+02	8.2963
0	4	4	1.10095	6.907 E+02	4.844 E+02	3.0409
4	3	2	1.09737	1.706 E+02	1.318 E+02	3.0026
4	3	2	1.09737	1.248 E+02	1.318 E+02	1.0956

$Ca_{0.7}Sm_{0.3}TiO_3$

1	0	1	3.83305	9.568 E+02	4.388 E+02	1.5134
1	1	1	3.42760	1.859 E+03	4.755 E+02	4.9673
1	1	1	3.42760	1.881 E+03	4.755 E+02	2.5036
2	0	0	2.72335	6.736 E+04	6.807 E+04	32.7537
2	0	0	2.72335	6.468 E+04	6.807 E+04	15.6740
1	2	1	2.70883	5.177 E+04	5.465 E+04	100.00
1	2	1	2.70883	5.160 E+04	5.465 E+04	49.6741
0	0	2	2.69758	5.298 E+04	5.696 E+04	25.4472
0	0	2	2.69758	5.276 E+04	5.696 E+04	12.6309
2	1	0	2.56590	4.505 E+03	5.781 E+03	4.0632
2	1	0	2.56590	6.907 E+03	5.781 E+03	3.1050
2	0	1	2.43117	1.552 E+03	1.159 E+03	1.3106

1	0	2	2.43735	1.432 E+03	1.827 E+03	1.2011
2	1	1	2.31719	1.700 E+03	5.542 E+03	2.7128
2	1	1	2.31719	1.775 E+03	5.542 E+03	1.4127
0	3	1	2.30727	4.915 E+03	1.858 E+03	3.9023
0	3	1	2.30727	7.083 E+03	1.858 E+03	2.8040
2	2	0	2.21921	1.282 E+04	1.151 E+04	9.7345
2	2	0	2.21921	1.401 E+04	1.151 E+04	5.3036
0	2	2	2.20520	1.155 E+04	9.684 E+03	8.7118
0	2	2	2.20520	1.160 E+04	9.684 E+03	4.3621
1	3	1	2.12452	1.091 E+03	9.873 E+02	1.5783
2	2	1	2.05236	1.013 E+03	1.212 E+03	1.4120
2	0	2	1.91652	6.104 E+04	5.279 E+04	39.6157
2	0	2	1.91652	5.868 E+04	5.279 E+04	18.9966
0	4	0	1.91434	8.438 E+04	7.671 E+04	27.3520
0	4	0	1.91434	8.389 E+04	7.671 E+04	13.5629
2	3	0	1.86234	2.390 E+03	1.421 E+04	1.5071
2	1	2	1.85918	3.372 E+03	2.025 E+03	4.2445
2	1	2	1.85918	4.708 E+03	2.025 E+03	2.9561
2	3	1	1.76041	1.999 E+03	1.593 E+01	2.3852
0	1	3	1.75075	2.551 E+03	1.521 E+03	1.5139
3	1	1	1.67888	5.302 E+03	2.359 E+03	6.0451
3	1	1	1.67888	4.337 E+03	2.359 E+03	2.4666
3	2	1	1.56952	1.489 E+04	1.267 E+04	15.950
3	2	1	1.56952	1.368 E+04	1.267 E+04	7.3112
2	4	0	1.56613	3.124 E+04	2.924 E+04	16.6961
2	4	0	1.56613	3.039 E+04	2.924 E+04	8.1041

0	4	2	1.56118	2.989 E+04	2.821 E+04	15.9299
1	2	3	1.55961	1.538 E+04	1.376 E+04	16.3839
0	4	2	1.56118	3.398 E+04	2.821 E+04	9.0341
1	2	3	1.55961	1.830 E+04	1.376 E+04	9.7243
2	3	2	1.53259	3.381 E+03	1.923 E+03	3.5444
3	3	1	1.42680	2.525 E+03	1.341 E+03	2.4820
3	3	1	1.42680	2.224 E+03	1.341 E+03	1.0906
4	0	0	1.36167	2.383 E+04	1.707 E+04	5.6173
4	0	0	1.36167	2.128 E+04	1.707 E+04	2.5025
2	4	2	1.35442	2.089 E+04	2.144 E+04	19.6052
2	4	2	1.35442	1.911 E+04	2.144 E+04	8.9479
4	1	0	1.34064	2.958 E+03	3.655 E+03	1.3756
0	0	4	1.34879	1.298 E+04	1.426 E+04	3.0349
0	0	4	1.34879	1.595 E+04	1.426 E+04	1.8604
3	1	3	1.26026	2.519 E+03	8.272 E+02	1.1060
4	0	2	1.21559	7.886 E+03	1.088 E+04	3.3579
4	0	2	1.21559	9.629 E+03	1.088 E+04	2.0453
3	2	3	1.21198	7.017 E+03	7.710 E+03	5.9591
1	6	1	1.21087	7.391 E+03	7.738 E+03	6.2717
3	2	3	1.21198	7.168 E+03	7.710 E+03	3.0369
2	0	4	1.20867	9.780 E+03	1.065 E+04	4.1423
1	6	1	1.21087	6.819 E+03	7.738 E+03	2.8865
2	0	4	1.20867	1.065 E+03	1.065 E+04	1.9135
4	1	2	1.20055	1.791 E+03	2.574 E+03	1.5075
4	1	2	1.20055	3.131 E+03	2.574 E+03	1.3151
3	5	1	1.14400	2.058 E+03	6.754 E+02	1.6559

3	3	3	1.14253	1.941 E+03	6.590 E+02	1.5601
4	4	0	1.10960	8.494 E+03	8.124 E+03	3.3182
4	4	0	1.10960	1.127 E+03	8.124 E+03	2.1960
0	4	4	1.10260	1.010 E+04	7.478 E+03	3.9193
0	4	4	1.10260	1.004 E+04	7.478 E+03	1.9437
4	3	2	1.09748	2.814 E+03	1.555 E+03	2.1747
4	3	2	1.09748	3.275 E+03	1.555 E+03	1.2623

$Ca_{0.6}Sm_{0.4}TiO_3$

1	0	1	3.84438	6.226 E+02	5.895 E+02	2.2056
1	0	1	3.84438	9.692 E+02	5.895 E+02	1.7095
1	1	1	3.43723	8.814 E+02	1.644 E+02	5.1902
1	1	1	3.43723	1.115 E+03	1.644 E+02	3.2710
2	0	0	2.70656	4.058 E+04	4.270 E+04	41.2985
0	0	2	2.73037	1.470 E+04	1.557 E+04	15.1512
1	2	1	2.71584	2.444 E+04	2.617 E+04	100.00
2	0	0	2.70656	3.869 E+04	4.270 E+04	19.6178
0	0	2	2.73037	1.465 E+04	1.557 E+04	15.1512
1	2	1	2.71584	2.444 E+04	2.617 E+04	50.8394
2	1	0	2.55248	7.112 E+03	3.443 E+03	13.3020
2	1	0	2.55248	5.540 E+03	3.443 E+03	5.1632
1	0	2	2.43781	8.344 E+02	7.251 E+02	1.4631
2	0	1	2.42504	6.561 E+02	7.900 E+02	1.1421
2	0	1	2.42504	1.247 E+03	7.900 E+02	1.0818
2	1	1	2.31234	1.057 E+03	3.024 E+02	3.4497
2	1	1	2.31234	8.978 E+03	3.024 E+02	1.4598
0	3	1	2.31653	1.253 E+03	1.122 E+03	2.0448

0	3	1	2.31653	3.059 E+03	1.122 E+03	2.4932
2	2	0	2.21173	7.349 E+03	5.707 E+03	11.2985
2	2	0	2.21173	6.528 E+03	5.707 E+03	5.0020
0	2	2	2.22466	4.015 E+03	2.529 E+03	6.2203
0	2	2	2.22466	3.973 E+03	2.529 E+03	3.0681
1	3	1	2.12971	1.088 E+03	5.292 E+02	3.1864
1	3	1	2.12971	8.070 E+03	5.292 E+02	3.1864
2	2	1	2.04997	1.633 E+03	8.498 E+03	4.5552
2	2	1	2.04997	7.558 E+03	8.498 E+03	1.0507
1	2	2	2.05767	5.825 E+02	4.665 E+02	1.6323
2	0	2	1.92219	2.542 E+04	2.285 E+04	32.7481
2	0	2	1.92219	2.504 E+04	2.285 E+04	16.0776
0	4	0	1.91859	6.060 E+04	5.590 E+04	38.9447
0	4	0	1.91859	5.880 E+04	5.590 E+04	18.8364
2	1	2	1.86459	1.397 E+03	7.745 E+02	3.4713
2	1	2	1.86459	1.024 E+03	7.745 E+02	1.2687
2	2	2	1.71861	8.237 E+02	3.840 E+02	1.8617
1	1	3	1.68330	2.893 E+03	1.345 E+03	6.3888
3	1	1	1.67211	1.744 E+03	1.413 E+03	1.9055
1	1	3	1.68330	1.312 E+03	1.345 E+03	1.4450
2	4	0	1.56522	2.704 E+04	2.267 E+04	27.5686
2	4	0	1.56522	3.101 E+04	2.267 E+04	15.7657
1	2	3	1.57357	7.002 E+04	5.599 E+04	14.3601
3	2	1	1.56441	1.145 E+03	9.183 E+03	23.3371
0	4	2	1.56979	1.682 E+04	1.531 E+04	17.2049
1	2	3	1.57357	6.223 E+03	5.599 E+03	6.3644

3	2	1	1.56441	1.265 E+04	9.183 E+03	12.8549
0	4	2	1.56979	1.639 E+04	1.531 E+04	8.3600
2	3	2	1.53671	1.366 E+03	1.038 E+03	2.6711
2	1	3	1.48201	1.053 E+03	3.637 E+03	2.0265
2	1	3	1.48201	1.346 E+03	3.637 E+03	1.2916
3	1	2	1.47721	1.067 E+03	1.509 E+01	1.0203
3	3	1	1.42351	2.121 E+03	1.009 E+03	3.9135
3	3	1	1.42351	1.884 E+03	1.009 E+03	1.7341
3	2	2	1.40138	5.812 E+02	3.835 E+01	1.0556
0	0	4	1.36519	1.485 E+04	5.602 E+03	6.5665
0	0	4	1.36519	8.206 E+04	5.602 E+03	1.8094
2	4	2	1.35792	1.593 E+04	1.455 E+04	28.0248
2	4	2	1.35792	1.521 E+04	1.455 E+04	13.3469
4	0	0	1.35328	1.655 E+04	1.495 E+04	7.2529
4	0	0	1.35328	2.438 E+04	1.495 E+04	5.3274
2	5	0	1.33513	1.385 E+03	1.720 E+03	1.1972
4	0	1	1.31355	1.325 E+03	4.161 E+02	1.1272
2	5	1	1.29693	7.879 E+02	2.491 E+02	1.3231
2	5	1	1.29693	1.299 E+03	2.491 E+02	1.0883
0	2	4	1.28621	1.377 E+03	4.352 E+02	1.1467
1	2	4	1.25137	1.332 E+03	8.038 E+01	2.1584
1	2	4	1.25137	1.596 E+03	8.038 E+01	1.2901
4	2	1	1.24275	1.099 E+03	6.291 E+01	1.7688
4	2	1	1.24275	1.875 E+03	6.291 E+01	1.5054
2	0	4	1.21891	4.730 E+03	5.373 E+03	3.7337
3	2	3	1.21547	4.921 E+03	5.461 E+03	7.7470

1	6	1	1.21365	7.845 E+03	8.206 E+03		12.3324
2	0	4	1.21891	4.668 E+03	5.373 E+03		1.8378
4	0	2	1.21252	8.360 E+03	8.524 E+03		6.5651
3	2	3	1.21547	5.376 E+03	5.461 E+03		4.2216
1	6	1	1.21365	8.613 E+03	8.206 E+03		6.7533
4	0	2	1.21252	1.035 E+04	8.524 E+03		4.0554
2	1	4	1.20382	1.038 E+03	2.418 E+02		1.6189
2	5	2	1.19941	7.640 E+02	3.841 E+02		1.1870
4	3	0	1.19621	3.736 E+03	3.353 E+03		2.8949
4	1	2	1.19766	2.575 E+03	1.997 E+03		3.9947
4	3	0	1.19621	3.502 E+03	3.353 E+03		1.3533
4	1	2	1.19766	1.776 E+03	1.997 E+03		1.3746
1	3	4	1.17566	1.087 E+03	2.395 E+01		1.6557
0	5	3	1.17339	5.490 E+03	3.509 E+02		4.1731
1	5	3	1.14676	7.178 E+02	1.851 E+02		1.0667
3	3	3	1.14574	2.056 E+03	5.207 E+02		3.0533
3	5	1	1.14320	2.524 E+03	7.809 E+02		3.7401
3	3	3	1.14574	1.663 E+03	5.207 E+02		1.2318
2	6	1	1.13134	2.328 E+03	3.773 E+01		1.7026
0	4	4	1.11233	7.429 E+03	4.563 E+03		5.3551
4	4	0	1.10586	9.529 E+03	8.798 E+03		6.8289
0	4	4	1.11233	6.462 E+03	4.563 E+03		2.3235
4	4	0	1.10586	1.084 E+04	8.798 E+03		3.8734
4	3	2	1.09567	1.499 E+03	1.395 E+03		2.1292

- The reflection selected from the crystallographic information framework output of the final cycle of the refinement

- Intensities less than 1% were omitted

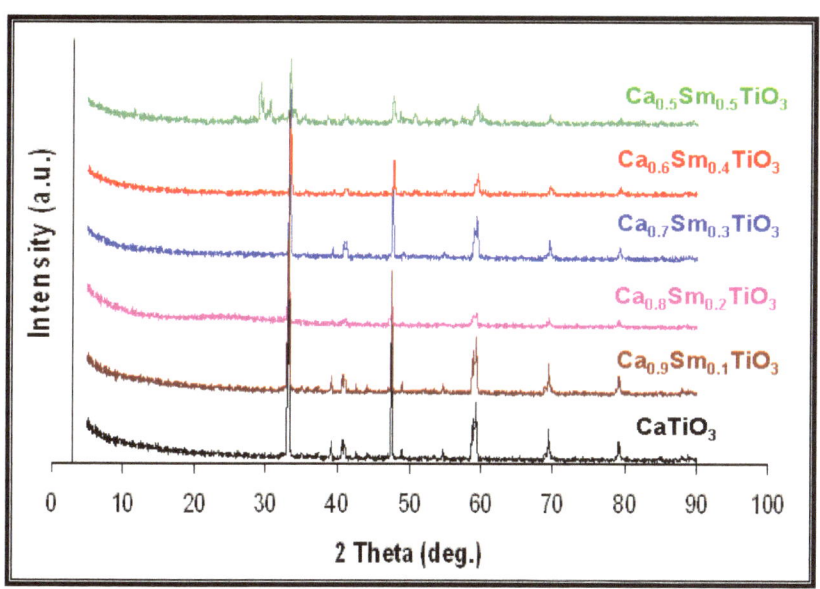

Figure 3.39 Powder XRD pattern of $Ca_{1-x}Sm_xTiO_3$ (x = 0.1-0.5) ceramic samples

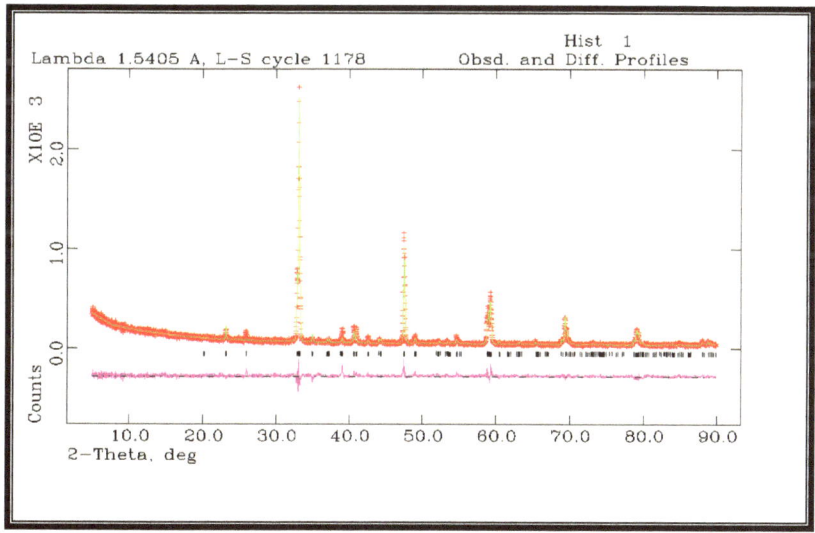

Figure 3.40 Rietveld refinement plot for $Ca_{0.9}Sm_{0.1}TiO_3$ ceramic sample showing observed (+), calculated (continuous line), and difference (lower) curves. The vertical bars denote Bragg reflections of the crystalline phases.

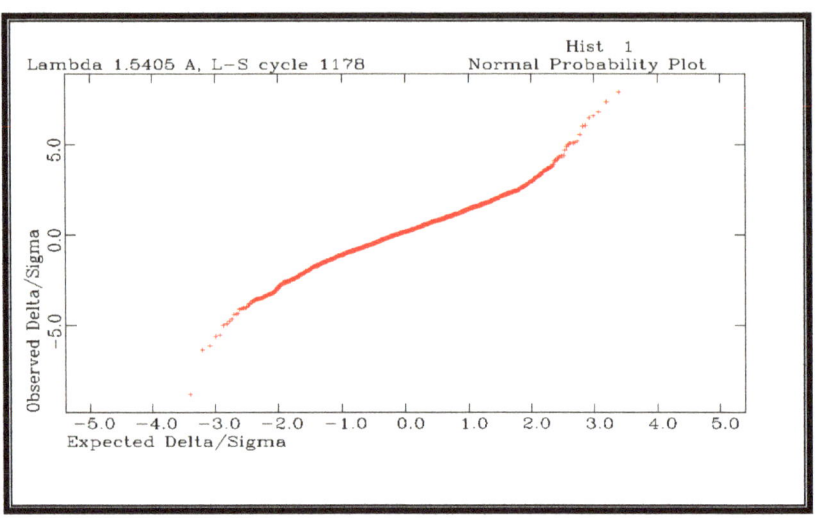

Figure 3.41 Probability plot between I_0 - I_C for $Ca_{0.9}Sm_{0.1}TiO_3$ ceramic sample

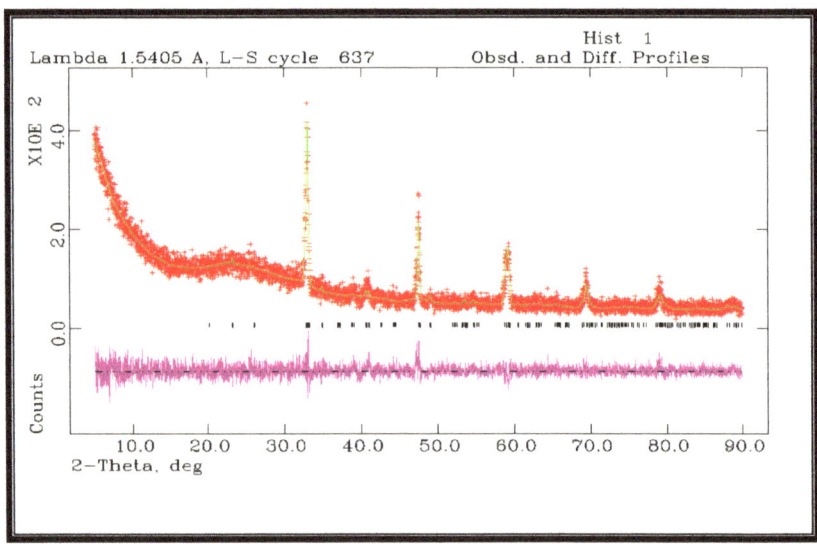

Figure 3.42 Rietveld refinement pattern of $Ca_{0.8}Sm_{0.2}TiO_3$. The measured intensity data and the differences between observed and calculated values are given in the *upper* and in the *lower plots*, respectively

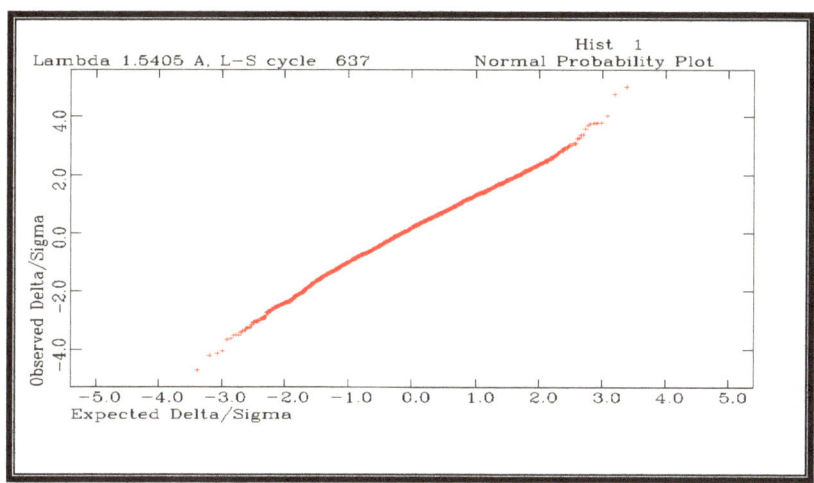

Figure 3.43 Probability plot between $I_0 - I_C$ for polycrystalline $Ca_{0.8}Sm_{0.2}TiO_3$ ceramic sample

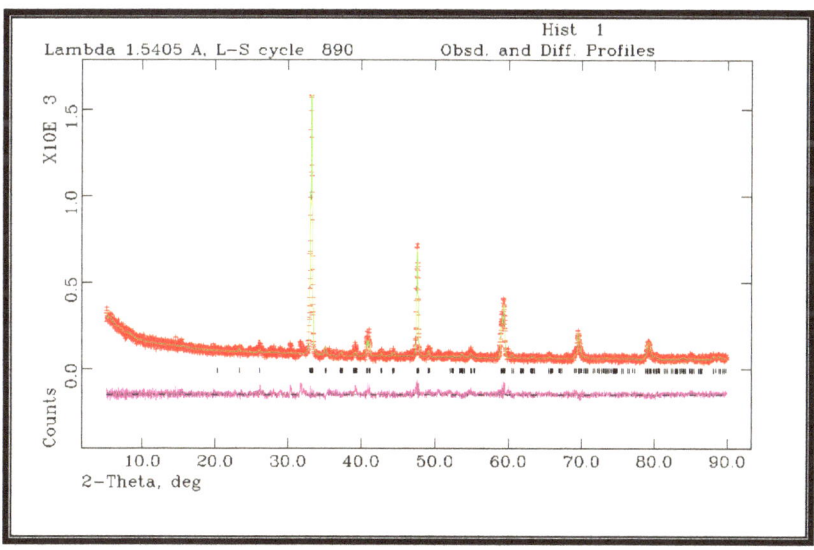

Figure 3.44 Rietveld refinement pattern of $Ca_{0.7}Sm_{0.3}TiO_3$. The measured intensity data and the differences between observed and calculated values are given in the *upper* and in the *lower plots*, respectively

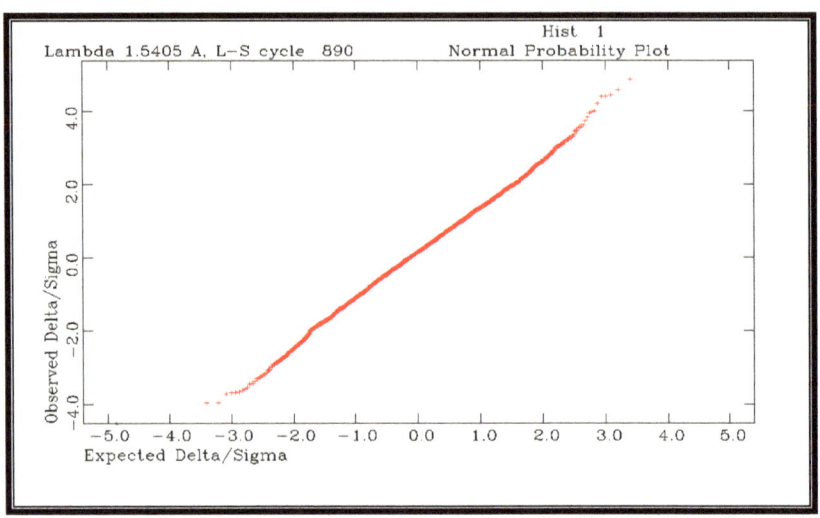

Figure 3.45 Probability plot between I_0 - I_C for polycrystalline $Ca_{0.7}Sm_{0.3}TiO_3$ ceramic sample

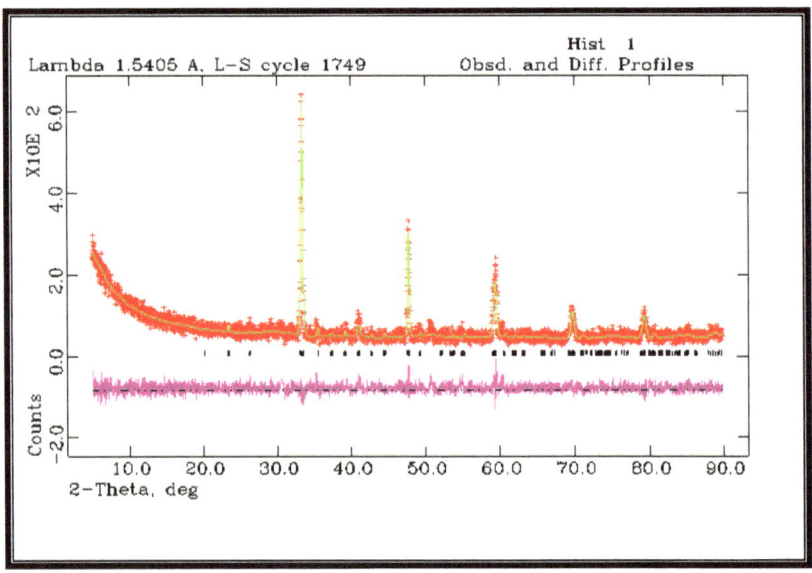

Figure 3.46 Rietveld refinement pattern of $Ca_{0.6}Sm_{0.4}TiO_3$. The measured intensity data and the differences between observed and calculated values are given in the *upper* and in the *lower plots*, respectively

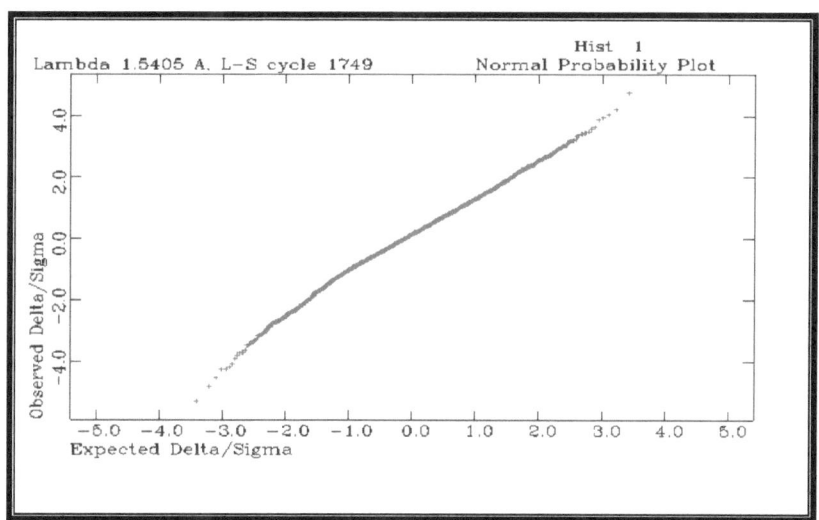

Figure 3.47 Probability plot between I_0 - I_C for polycrystalline $Ca_{0.6}Sm_{0.4}TiO_3$ ceramic sample

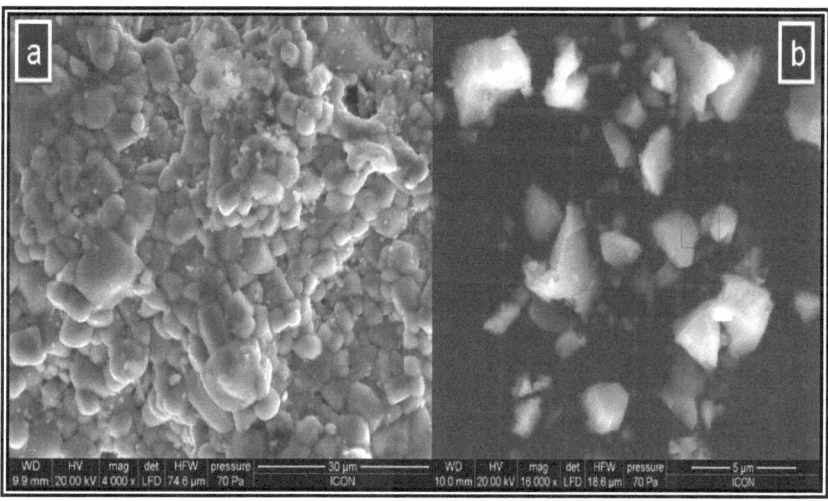

Figures 3.48 (a) Scanning electron micrographs of $Ca_{0.9}Sm_{0.1}TiO_3$ ceramic powder and (b) SEM of $Ca_{0.9}Sm_{0.1}TiO_3$ showing location of elemental analysis

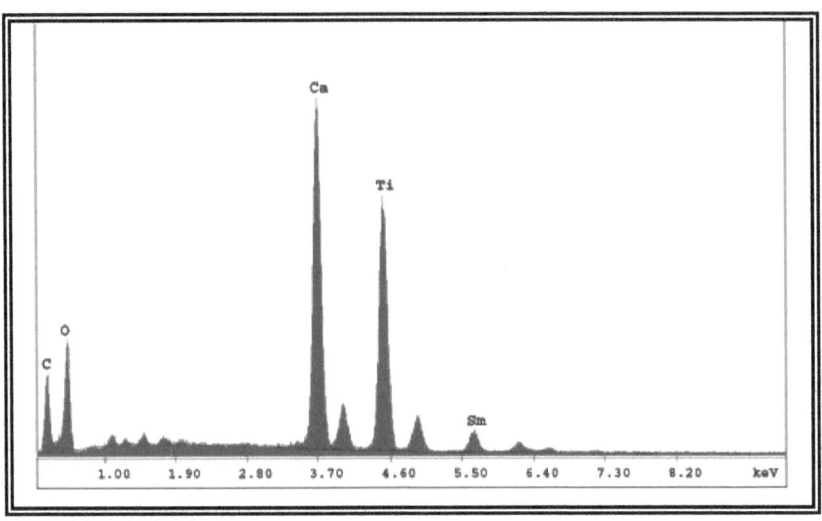

Figure 3.49 EDAX spectrum of polycrystalline mono phase of $Ca_{0.9}Sm_{0.1}TiO_3$ ceramic

Chapter 4

Outcomes of the Research Work

4.1 Metal Substituted Sodium Zirconium Phosphates (NZP)

4.1.1 Metal Strontium substituted sodium zirconium phosphate:

$Na_{1-x}Sr_{x/2}Zr_2P_3O_{12}$ (where x = 0.1-1.0)

Rietveld refinement and crystallographic model of the phases

The powder XRD data showed that mono phases of composition $Na_{1-x}Sr_{x/2}Zr_2P_3O_{12}$ (x = 0.1 to 0.7) are isostructural to $NaZr_2(PO_4)_3$ which is confirmed from the diffraction patterns (Fig. 3.1). SrNZP crystallizes in the rhombohedral system (Space group R-3c). The conditions for the rhombohedral lattice: (i) -h+k+l=3n (ii) when h=0, l=2n and (iii) when k=0, l=2n have been verified for all reflections between 2θ =10°-120°. The intensity and positions of the diffraction pattern match with the characteristic pattern of parent compound sodium zirconium phosphate, which gives several prominent reflections between 2θ=13.98°- 46.47° [1].

The General Structure Analysis System (GSAS) [2] program with the EXPGUI [3] graphical user interface was used for Rietveld [4] analysis of the X-ray powder diffraction data. High quality powder diffraction data in combination with the Rietveld method allows refinement of a structural model (atomic coordinates, site occupancies and atomic displacement parameters) as well as profile parameters (lattice constants, peak shape, sample height, instrument parameters and background). Assuming that SrNZP belongs to the NASICON (Sodium super ionic conductor) family [5], Zr, P and O atoms are in the 12c, 18e and 36f Wyckoff positions respectively of the R-3c space group. In SrNZP phases, Na atoms were assumed to occupy the M_1 (6b) site while M_2 site (18e) remains vacant. The occupancies of Na and Sr atoms have been constrained according to their theoretical molar ratios. The structure refinement leads to rather good agreement between the experimental and calculated XRD pattern (Fig. 3.2, 3.4, 3.6, 3.8 & 3.10) and yields acceptable reliability factors: RF^2, Rp and Rwp [6]. The normal Probability plots for the histogram gives nearly a linear relationship indicating that the *Io* and *Ic* values for the most part are normally distributed (Fig. 3.3, 3.5, 3.7, 3.9 & 3.11). The lattice parameters are close to the corresponding values for un-substituted NZP unit cell [7]. The cell parameters of the specimens register slight increase in *a* direction (Table 3.1). Simultaneously, the structure shows a little contraction along *c* direction.Due to angular distortions as a result of the coupled rotation of ZrO_6 and PO_4 polyhedrons [8]. Alteration in lattice parameters shows that the network modifies its dimensions to accommodate the cations occupying M_1 site without breaking the bonds. The basic framework of NZP accepts the cations of different sizes and oxidation states to form solid solutions but at the same time retaining the overall geometry unchanged. The final atomic coordinates and isotropic thermal parameters (Table 3.2), inter-atomic distances (Table 3.3) and bond angles (Table 3.4) are extracted from the crystal information file (CIF) prepared after final cycle of refinement. Selected h, k, l values, d-spacing and intensity data along with observed and calculated structure factors have been listed in table 3.5.The refinement leads to acceptable Zr-O, P-O bond distances. Zr atoms are displaced from the center of the octahedron due to the $Na^+–Zr^{4+}$ repulsions. Consequently the Zr–O(2) distances (2.0708, 2.0697, 2.073, 2.0709 & 2.0750 Å), neighboring sodium Na(1), is slightly greater than the Zr–O(1) distances (2.0276, 2.0265, 2.0304, 2.0277 & 2.0306 Å), however, average Zr–O distances are smaller than the values calculated from the ionic radii data (2.12 Å) [9]. The O–Zr–O angles vary between 84.76° and 175.96°. The angles implying the shortest bonds are superior to those involving the longest ones due to O-O repulsions which are stronger for O(1)-O(1) than for O(1)-O(2).

The P-O distances (in pairs) 1.547, 1.546, 1.549, 1.547 and 1.553 Å are close to those found in NASICON type phosphates [10-14]. The O-P-O angles vary between 106.18°-111.91°. Fig. 4.1 shows the PLATON projection of the molecular structure depicting the inter linking of ZrO_6 and PO_4 through a bridge oxygen atom.

The Oak Ridge Thermal Ellipsoid Plot Program (hereafter ORTEP) view generated by the refined structural data of $Na_{0.9}Sr_{0.05}Zr_2P_3O_{12}$ shows that, the Zr–O bonds are in three pairs resulting in six coordination of zirconium. The Zr–O bond distances occur in two sets: 2.027 and 2.070 Å for Sr substituted NZP. The PO_4 tetrahedra are nearly regular. The P–O distances are 1.51826(20) and 1.54751(20) for Sr substituted NZP. The planer O–Zr–O angles are different from those connecting the apex oxygen atoms of the octahedron. The O1–Zr–O2 bond angle is 175.961° for Sr substituted NZP, indicating that the ZrO_6 octahedra are slightly tilted. The other O–Zr–O angle values occur in triplicate (Fig. 4.2). Figure 4.3 & 4.4 shows the ball & stick and diamond view respectively of $Na_{0.9}Sr_{0.05}Zr_2P_3O_{12}$ ceramic phase, which illustrate that the structure is based on a 3D framework built from ZrO_6 octahedra and corner sharing PO_4 tetrahedra.

Polyhedral distortions of ZrO_6 and PO_4

From the individual bond lengths of the metal-oxygen polyhedra reported by various authors the polyhedral distortion can be calculated by the following equation $\Delta = 1/n \, \Sigma\{(R_i - R_m)/(R_m)\}^2$ where R_i= individual bond length, R_m= average bond length, n is the number of M-O bonds [15-18]. The calculated distortions in ZrO_6 octahedra and PO_4 tetrahedra have been summarized in Table 3.3.

Particle size distribution

The X-ray data of the title phase was used for the estimation of particle size using Scherrer's formula [19-20], $t = 0.9\lambda/\beta \cos\theta$ where t is the crystallite size as measured perpendicular to the reflecting plane, K=0.9 the Scherrer's constant, λ the wavelength of X-ray radiation, β = full width at the half intensity of maxima (FWHM) measured in radians and θ = the Bragg angle. The (h k l) values corresponding to prominent reflections, their half width and respective particle size are listed in Table 4.1. The particle size distribution along the prominent reflecting planes and full width at the half intensity of maxima mentioned in the table varies between 21.08-195.8 nm.

SEM and EDX analysis

The microstructure of the title phase has been examined by SEM and EDAX analysis of the specimen. The evolution of SrNZP phase can be seen clearly in the electron micrographs of the ceramic sample (Fig. 3.12-3.17).

Within the limits of experimental error, the EDAX analytical data on atomic and wt % of Na, Zr, P and Sr are found agreeable with their corresponding expected molar ratios. The EDAX spectrum shows that strontium crystallochemically enters into the NZP matrix.

IR Analysis

In the IR spectra (Fig. 4.5) of the $Na_{1-x}Sr_{x/2}Zr_2P_3O_{12}$ (x = 0.1-1.0) composition; Table 4.2 summarizes the vibrational frequencies of $(PO_4)^{3-}$ ions in these phases. Bands in the region of 1280 cm^{-1} to 1020 cm^{-1} are assigned to the stretching asymmetric vibrations υ_3, and bands in the region of 980cm^{-1} to 915 cm^{-1} correspond to the stretching symmetric vibrations υ_1 of the $(PO_4)^{3-}$ ions. Bands in the 670-400 cm^{-1} assigned to the bending vibrations υ_4 and υ_2.

Based on the analysis of the presented IR spectra, we assumed that phosphate of the $Na_{1-x}Sr_{x/2}Zr_2P_3O_{12}$ (x = 0.1-1.0) composition to be characterized by the *R-3c* space group and the prepared samples can be attributed to the orthophosphate class [21-23].

Table 4.1 Distribution of particle size (nm) along with prominent reflecting planes of $Na_{1-x}Sr_{x/2}Zr_2P_3O_{12}$ (x = 0.1-1.0) ceramic sample

h k l	FWHM (2θ)	Particle size (nm)
$Na_{0.9}Sr_{0.05}Zr_2P_3O_{12}$		
1 0 -2	0.1428	72.13
1 1 6	0.0836	105.43
2 1 1	0.1632	80.62
2 0 8	0.1224	124.6
2 2 0	0.1224	114.21
1 1 9	0.4080	80.62
2 1 -8	0.1428	76.14
3 1 -4	0.2676	31.87
2 0 -10	0.1004	80.62
4 0 -2	0.1632	97.9

2 1 10	0.0816	124.6
3 1 8	0.3264	28.55
3 2 4	0.1224	76.14
1 0 -14	0.1224	91.37
4 1 6	0.0816	124.6
4 1 -6	0.1224	97.9
2 0 14	0.2448	39.16
3 0 0	0.1632	80.62
1 1 15	0.1224	91.37
2 1 -14	0.1632	65.26
3 3 6	0.1224	76.14
3 1 14	0.1224	80.62
6 0 0	0.1224	105.43
4 2 -10	0.1632	97.9
3 2 -14	0.1224	97.9
5 2 -6	0.1632	91.37
5 0 14	0.1224	124.6
4 3 10	0.3264	29.79
6 1 -10	0.2448	40.31
7 1 0	0.2448	41.53
5 3 -10	0.1632	76.14
5 3 14	0.2448	52.71

Na$_{0.7}$Sr$_{0.15}$Zr$_2$P$_3$O$_{12}$

h k l				
1 0 -2		0.1338		62.6
1 1 3		0.1171		72.13
1 1 6		0.0836		105.43
2 1 1		0.0816		124.6
2 1 4		0.0816		114.21
3 0 0		0.1020		91.37
2 0 8		0.4015		21.08
1 1 9		0.1224		80.62
2 1 -8		0.1020		97.9
3 1 -4		0.1004		85.66
2 0 -10		0.0612		152.28
2 2 6		0.0612		152.28
3 1 8		0.0816		114.21
3 2 4		0.0669		137.06
4 1 0		0.0612		152.28
4 0 -8		0.1224		97.9
4 1 -6		0.0816		171.32
2 0 14		0.0816		124.6
5 0 -4		0.0612		171.3
3 3 0		0.1224		80.62
4 0 10		0.1020		97.9
2 1 -14		0.0816		114.21
3 2 -10		0.1224		91.37
5 1 4		0.1020		105.43
3 2 -14		0.1224		72.13

5 2 6	0.0816	124.6
4 3 -8	0.1224	80.62
5 1 -14	0.0816	124.6

$Na_{0.5}Sr_{0.25}Zr_2P_3O_{12}$

1 0 -2	0.1020	97.9
1 0 4	0.1020	97.9
1 1 6	0.1171	72.13
2 1 1	0.1224	76.14
2 1 4	0.1004	85.66
3 0 0	0.0612	152.28
2 0 8	0.1020	105.43
1 1 9	0.0816	137.06
3 0 -6	0.0816	137.06
2 1 -8	0.0816	124.6
2 1 10	0.1020	91.37
4 1 10	0.1338	65.26
1 0 -14	0.0816	80.62
4 0 -8	0.1224	80.62
3 1 -10	0.1020	97.9
4 1 6	0.0836	114.21
4 1 -6	0.1224	85.66
5 0 -4	0.1004	91.33
3 3 0	0.1632	76.14
4 0 10	0.1224	80.62
2 1 -14	0.1224	76.14
3 2 10	0.1673	52.71

6 0 0	0.1224	85.66
4 3 10	0.1224	91.37

Na$_{0.3}$Sr$_{0.35}$Zr$_2$P$_3$O$_{12}$

1 0 -2	0.1171	72.13
6 1 1	0.0816	114.21
1 1 6	0.0836	105.43
2 1 1	0.1338	62.3
2 1 4	0.0612	159.37
2 0 8	0.1632	59.59
1 1 9	0.0816	124.6
3 0 -6	0.0816	124.6
2 1 -8	0.0816	114.21
3 1 -4	0.1224	80.62
2 0 -10	0.0816	124.6
2 0 10	0.0836	105.4
3 1 8	0.0612	152.28
3 2 4	0.0836	105.4
4 1 0	0.0816	124.6
4 1 -6	0.1224	80.62
2 0 14	0.1020	97.9
5 0 -4	0.1224	97.9
3 3 0	0.0816	137.06
2 1 -14	0.1224	72.13
6 6 0	0.2448	36.08
3 2 -14	0.0612	171.32
4 3 10	0.0612	195.8

$Sr_{0.5}Zr_2P_3O_{12}$

1 0 -2	0.0816	114.21
1 0 4	0.1428	68.53
1 1 6	0.0669	137.06
2 1 1	0.1673	52.71
2 1 4	0.0669	152.28
3 0 0	0.1071	72.13
2 0 8	0.0612	95.8
1 1 9	0.1020	97.9
3 0 -6	0.1428	80.62
2 1 -8	0.0816	114.21
3 1 -4	0.0669	137.06
2 0 -10	0.1338	62.3
2 2 6	0.0612	152.28
4 0 -2	0.1428	68.53
3 1 -7	0.1224	85.66
3 1 8	0.0669	137.06
3 2 4	0.0669	137.06
1 0 -14	0.1338	62.3
3 1 -10	0.0612	152.28
2 0 -14	0.1428	65.26
2 0 14	0.1428	80.62
4 0 10	0.0612	171.32
5 1 4	0.1224	76.14
3 1 14	0.1632	62.3
6 0 0	0.1020	97.9

| 3 2 -14 | 0.2040 | 47.26 |
| 4 3 -8 | 0.2040 | 50.76 |

Table 4.2 Assignment (cm^{-1}) of IR bands for $Na_{1-x}Sr_{x/2}Zr_2P_3O_{12}$ (x = 0.1-1.0) ceramic samples

Compound	v_3 v_{as} (P-O)	v_1 v_a (P-O)	v_4 ∂(P-O)	v_2 (P-O)
$Na_{0.9}Sr_{0.05}Zr_2P_3O_{12}$	1022.31 1051.24 1060.88 1109.11 1201.69 1271.13	931.65 952.87 964.44	545.87 576.74 646.17 634.6 555.52	474.5
$Na_{0.7}Sr_{0.15}Zr_2P_3O_{12}$	1024.24 1037.74 1058.96 1112.96 1269.20	908.5 927.79 950.94 966.37 981.8	542.02 553.59 576.74 599.88 611.45	405.06 418.57 428.21 435.93 449.43
$Na_{0.5}Sr_{0.25}Zr_2P_3O_{12}$	1047.38 1060.88 1097.53 1138.04 1201.69 1273.06 1209.41	950.94 976.01	516.64 555.52 574.81 626.89 540.09	401.21 410.85 418.57 422.42 434.0
$Na_{0.3}Sr_{0.35}Zr_2P_3O_{12}$	1060.88 1095.6 1105.25 1128.39 1193.98 1203.62 1267.27	900.79 935.51 947.08 964.08 976.01	501.51 515.01 526.58 555.52 576.74	401.21 410.85 418.57 422.42 434.0
$Sr_{0.5}Zr_2P_3O_{12}$	1047.38 1060.88 1099.46 1186.26 1199.76	970.23 981.8	557.45 605.67 642.32	422.42 447.5 459.07 486.08

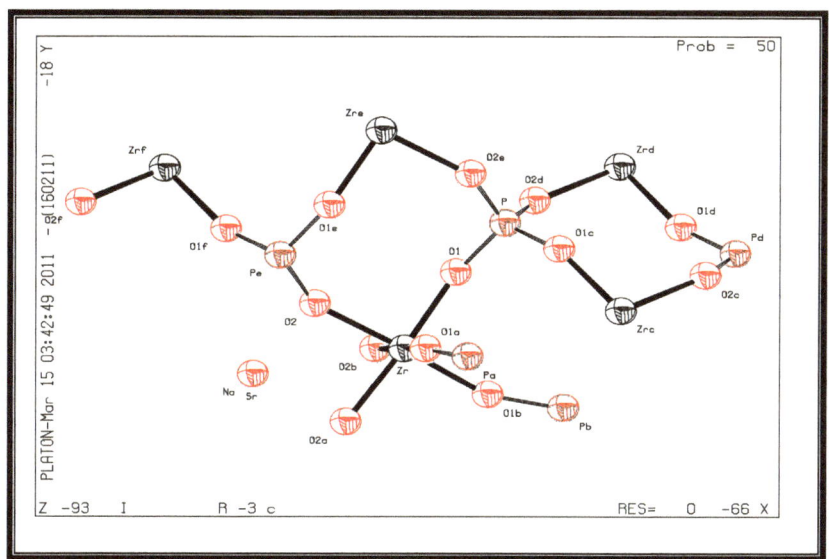

Fig. 4.1 PLATON view of molecular structure of $Na_{0.9}Sr_{0.05}Zr_2P_3O_{12}$ showing Zr coordination in ZrO_6 and P coordination in PO_4 polyhedron at 50% probability level

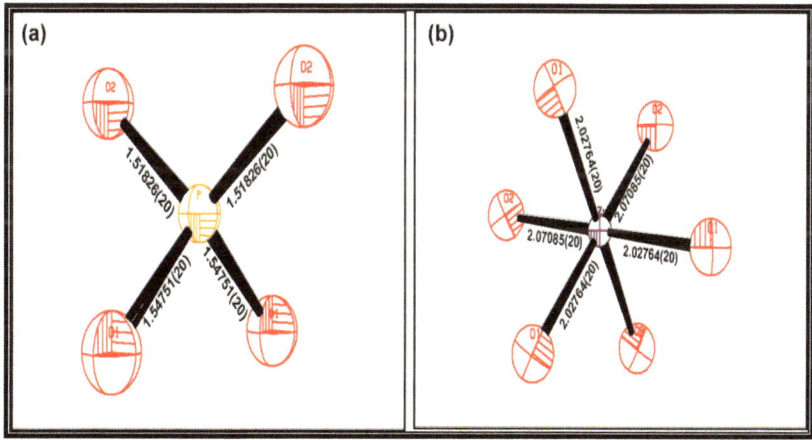

Fig. 4.2 ORTEP view of Zr coordination in ZrO_6 octahedra and P coordination in PO_4 tetrahedra in $Na_{0.9}Sr_{0.05}Zr_2P_3O_{12}$ normal to 001 plane

151

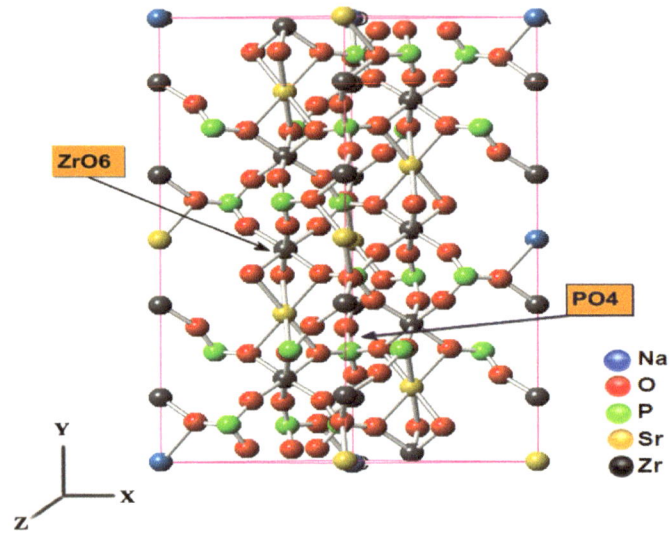

Fig.4.3 Ball and stick representation of $Na_{0.9}Sr_{0.05}Zr_2P_3O_{12}$ ceramic phase

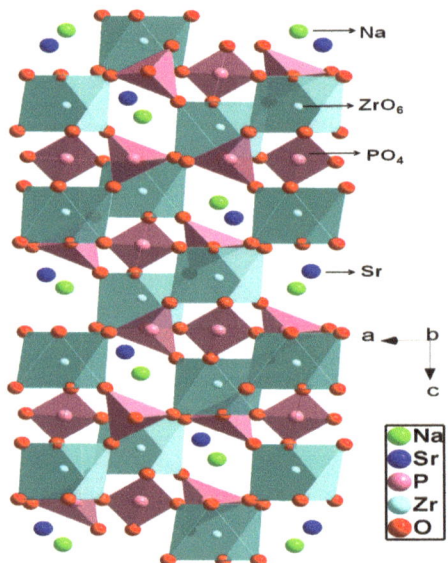

Fig.4.4 DIAMOND view of crystal structure of $Na_{0.9}Sr_{0.05}Zr_2P_3O_{12}$ ceramic phase showing substitution on Na (M-1) site

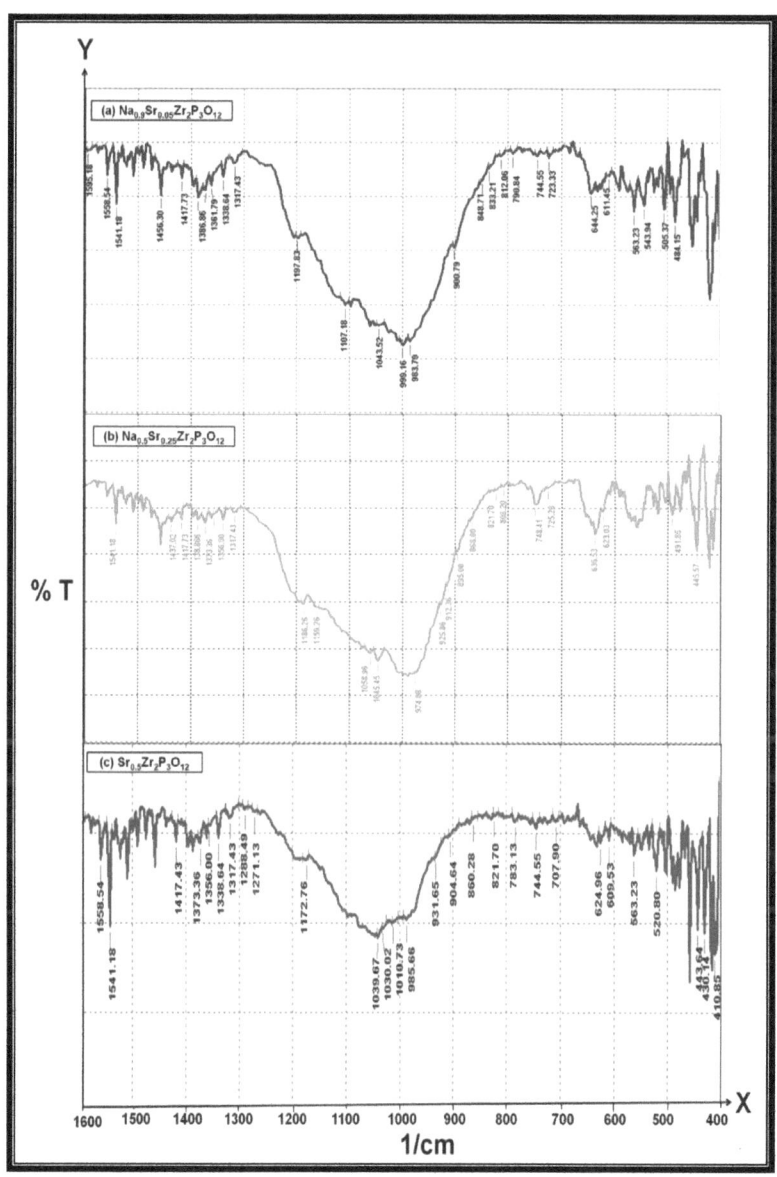

Figure 4.5 Infrared spectra of $Na_{1-x}Sr_{x/2}Zr_2P_3O_{12}$ (a) x = 0.1, (b) x = 0.3, (c) x = 0.5, (d) x = 0.7 & (e) x = 1.0

4.1.2 Cesium / Strontium substituted sodium zirconium phosphates:

$Na_{1-x}(Cs_{1.33}Sr_1)_xZr_2P_3O_{12}$ (where x = 0.1-1.0)

The intensities and positions of the diffraction patterns of the title phases matched fairly with the standard pattern of parent NZP which gives characteristic intense lines at 2θ = 13.981°, 19.467° 20.172°, 23.363°, 28.176° and 31.124° (Fig. 3.18). The Rietveld refinement of the step scan data was performed by the least square method using GSAS software, which is capable of handling and refining the diffraction data in a comprehensive manner. The phases crystallize in the rhombohedral system with space group R-3c. The conditions for the rhombohedral lattice: (i) -h + k + l = 3n, (ii) when h = 0, l = 2n and (iii) when k = 0, l = 2n have been verified for all reflections between 2θ = 10°-120°. The three-dimensional network structure of NZP consists of strongly bonded polyhedra, which imparts stability. The structure is flexible towards ionic substitutions on different available sites [24-25]. The lattice parameters are close to the corresponding values for un-substituted NZP unit cell. The cell parameters of the specimens register slight increase in *a* direction (Table 3.6). Simultaneously, the structure shows a little contraction along *c* direction. Due to angular distortions as a result of the coupled rotation of ZrO_6 and PO_4 polyhedrons. Alteration in lattice parameters shows that the network modifies its dimensions to accommodate the cations occupying M_1 site without breaking the bonds. The basic framework of NZP accepts the cations of different sizes and oxidation states to form solid solutions but at the same time retaining the overall geometry unchanged. In the basic NZP framework, Zr, P and O atoms are in the (12c), (18e) and (36f) Wyckoff positions, respectively of the R-3c space group. The initial atomic coordinates used for refinement of the crystal structure of the various compositions were derived from the parent compound. The structure refinement leads to rather good agreement between the experimental and calculated XRD pattern (Fig. 3.19, 3.21 & 3.23) and yields acceptable reliability factors: RF^2, Rp and Rwp. The normal Probability plots for the histogram gives nearly a linear relationship indicating that the *Io* and *Ic* values for the most part are normally distributed (Fig. 3.20, 3.22 & 3.24).

The final atomic coordinates and isotropic thermal parameters (Table 3.7), inter-atomic distances, polyhedral distortions (Table 3.8) and bond angles (Table 3.9) are extracted from the crystal information file (CIF) prepared after final cycle of refinement. Selected h, k, l values, d-spacing, and intensity data along with observed and calculated structure factors have been listed in table 3.10.The refinement leads to acceptable Zr-O, P-O bond distances. Fig. 4.6 shows the PLATON projection of the molecular structure depicting the inter linking of ZrO_6 and PO_4 through a bridge oxygen atom.

The Oak Ridge Thermal Ellipsoid Plot Program (hereafter ORTEP) view generated by the refined structural data of $Na_{0.5}Cs_{0.665}Sr_{0.5}Zr_2P_3O_{12}$ shows that, the Zr-O bonds are in three pairs resulting in six coordination of zirconium. The Zr-O bond distances occur in two sets: 2.047 and 2.076 Å for $Na_{0.5}Cs_{0.665}Sr_{0.5}Zr_2P_3O_{12}$ Cs/Sr substituted NZP. The PO_4 tetrahedra are nearly regular. The P-O distances are 1.53270(3) and 1.53421(2) for Cs/Sr substituted NZP. The planer O-Zr-O angles are different from those connecting the apex oxygen atoms of the octahedron. The O1-Zr-O2 bond angle is 174.214° for Cs/Sr substituted NZP, indicating that the ZrO_6 octahedra are slightly tilted. The other O-Zr-O angle values occur in triplicate (Fig. 4.7). Fig. 4.8 and 4.9 shows the ball & stick and diamond view respectively of $Na_{0.5}Cs_{0.665}Sr_{0.5}Zr_2P_3O_{12}$ ceramic phases, which illustrates that the structure is based on a 3D framework built from ZrO_6 octahedra and corner sharing PO_4 tetrahedra. The particle size along prominent reflecting planes has been estimated by Scherer's formula. It varies between 30 and 195.8 nm (Table 4.3).

The morphology and microstructure of the specimen have been examined by scanning electron microscopy. The evolution of solid mono phase has seen clearly in the electron micrograph of $Na_{1-x}(Cs_{1.33}Sr_1)_xZr_2P_3O_{12}$ (x = 0.1 to 1.0) ceramic powder (Fig. 3.25-3.30). The EDX spectrum confirms that Cesium and strontium are crystallochemically fixed in the NZP matrix.

The presence of orthophosphate anion was confirmed by their characteristic IR bands due to stretching and bending vibrations of P-O bonds of PO_4 tetrahedron [26-27]. In the spectra, two main regions can be identified in the range 1300–400 cm^{-1} that are attributed to the phosphate unit: the bands between 1250 and 900 cm^{-1} are ascribed to the stretching vibrations of the

PO$_4$ unit (v_1 and v_3 modes), the bands between 650 and 400 cm^{-1} are due to the deformation of the O-P-O angle (v_2 and v_4 modes) (Fig. 4.10).

Table 4.3 Distribution of particle size (nm) along with prominent reflecting planes of Na$_{1-x}$(Cs$_{1.33}$Sr$_1$)$_x$Zr$_2$P$_3$O$_{12}$ (x = 0.1-1.0) ceramic sample

h k l	FWHM(2θ)	Particle size(nm)
Na$_{0.8}$Cs$_{0.266}$Sr$_{0.2}$Zr$_2$P$_3$O$_{12}$		
1 0 -2	0.0816	124.6
1 1 3	0.0816	124.6
2 0 -4	0.0816	114.2
1 1 6	0.0669	137.06
2 1 1	0.0816	105.43
1 0 -8	0.0612	171.3
3 0 0	0.0816	114.21
1 1 9	0.1224	86.62
1 0 10	0.1020	80.62
3 0 -6	0.1224	97.9
2 1 -8	0.0816	114.2
3 1 -4	0.1224	72.13
2 0 -10	0.1224	72.13
2 2 6	0.1004	85.66
4 0 4	0.0816	171.35
2 1 10	0.0612	195.8
3 1 8	0.1224	85.66
3 2 4	0.0816	124.6
4 0 -8	0.1171	72.13
5 1 4	0.1020	114.21
6 0 0	0.1224	80.62

3 2 -14	0.1020	105.43
4 3 10	0.0612	171.52

$Na_{0.7}Cs_{0.399}Sr_{0.3}Zr_2P_3O_{12}$

4 3 -8	0.4015	41.23
3 0 -6	0.1673	70.24
3 1 8	0.1673	59.78
2 0 -10	0.1673	81.56
4 2 -4	0.1632	47.23
2 1 10	0.1338	78.25
1 1 6	0.1171	62.23
4 0 10	0.1632	59.56
2 1 -14	0.1224	60.23
3 2 4	0.1224	48.36
1 0 -2	0.1224	38.23
4 0 -8	0.1224	58.96
4 1 -6	0.1428	77.45
5 2 6	0.1632	89.25
2 0 14	0.1004	55.25
3 1 -10	0.1224	40.12
3 2 -14	0.1224	58.89
5 0 -4	0.1004	114.97
2 1 -8	0.1020	52.58
2 1 1	0.1020	98.23
5 2 0	0.1224	88.89
2 0 8	0.1020	116.78
5 1 4	0.0836	109.35

Na$_{0.5}$Cs$_{0.665}$Sr$_{0.5}$Zr$_2$P$_3$O$_{12}$

1 0 -2	0.1171	72.16
1 1 6	0.0669	137.06
2 1 1	0.1004	91.37
2 1 4	0.1338	65.26
2 0 8	0.1338	65.26
2 0 -10	0.1004	85.66
3 1 8	0.2676	34.26
3 2 4	0.2676	30.45
4 0 -8	0.2007	42.83
3 1 -10	0.1004	97.9
4 1 6	0.2007	33.42
2 1 -14	0.1171	72.13
4 2 -4	0.1004	97.9
3 2 10	0.1338	65.26
6 0 0	0.1673	52.71
0 0 18	0.2676	30.45
3 2 -14	0.1171	76.14
4 3 10	0.2342	38.07

Table 4.4 Assignment (cm^{-1}) of IR bands for Na$_{1-x}$(Cs$_{1.33}$Sr$_1$)$_x$Zr$_2$P$_3$O$_{12}$ (x = 0.1-1.0) ceramic samples

Compound	v_3 v_{as} (P-O)	v_1 v_a (P-O)	v_4 δ (P-O)	v_2 (P-O)
Na$_{0.8}$Cs$_{0.266}$Sr$_{0.2}$Zr$_2$P$_3$O$_{12}$	1030.02 1047.38 1060.88 1107.18 1124.54 1257.63	918.15 927.79 945.15 958.65	526.58 542.02 576.74	447.5 457.14
Na$_{0.7}$Cs$_{0.399}$Sr$_{0.3}$Zr$_2$P$_3$O$_{12}$	1028.09 1051.24 1111.23 1120.68 1134.18 1259.56	904.64 968.3	515.01 555.52 574.81 613.38 636.53	401.21 414.71 422.42 466.79
Na$_{0.5}$Cs$_{0.665}$Sr$_{0.5}$Zr$_2$P$_3$O$_{12}$	1058.96 1089.82 1107.18 1122.61 1259.56	939.36 989.52	518.87 547.8 594.1 621.1	406.99 418.57 428.21 478.36 497.65

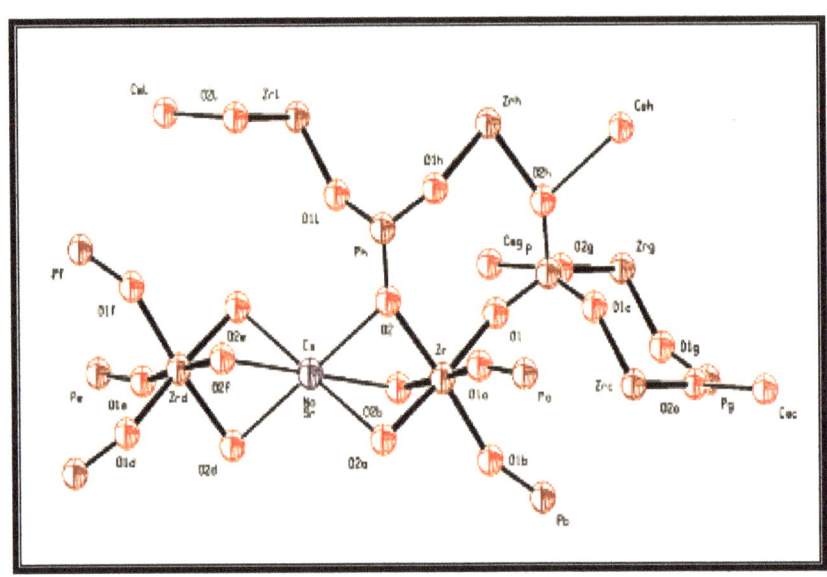

Fig. 4.6 PLATON view of molecular structure of $Na_{0.5}Cs_{0.665}Sr_{0.5}Zr_2P_3O_{12}$ showing Zr coordination in ZrO_6 and P coordination in PO_4 polyhedron at 50%probability level

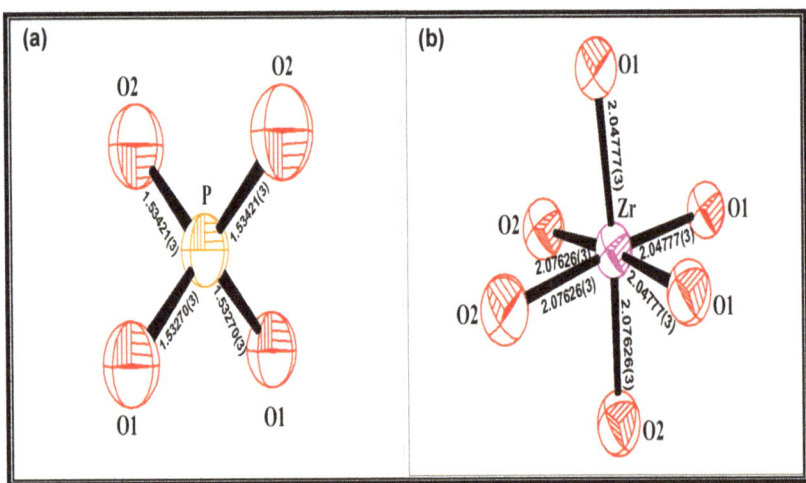

Fig.4.7 ORTEP view of Zr coordination in ZrO_6 octahedra and P coordination in PO_4 tetrahedra in $Na_{0.5}Cs_{0.665}Sr_{0.5}Zr_2P_3O_{12}$ normal to 001 plane

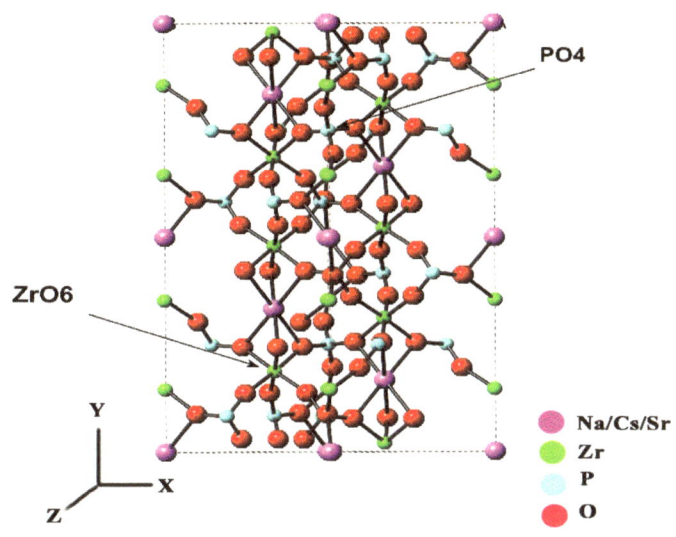

Fig.4.8 Ball and stick representation of $Na_{0.5}Cs_{0.665}Sr_{0.5}Zr_2P_3O_{12}$ ceramic phase.

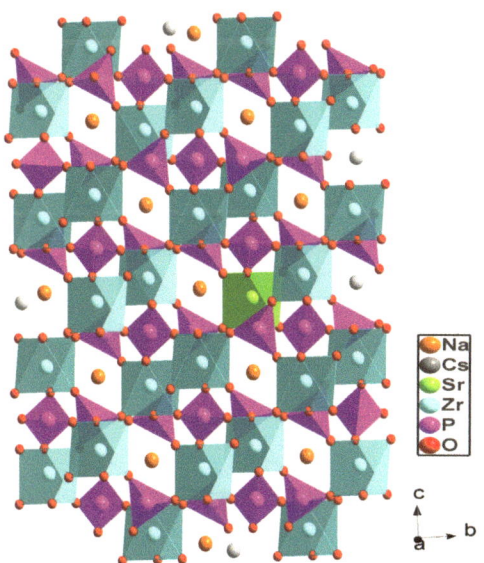

Fig.4.9 DIAMOND view of crystal structure of $Na_{0.5}Cs_{0.665}Sr_{0.5}Zr_2P_3O_{12}$ ceramic phase

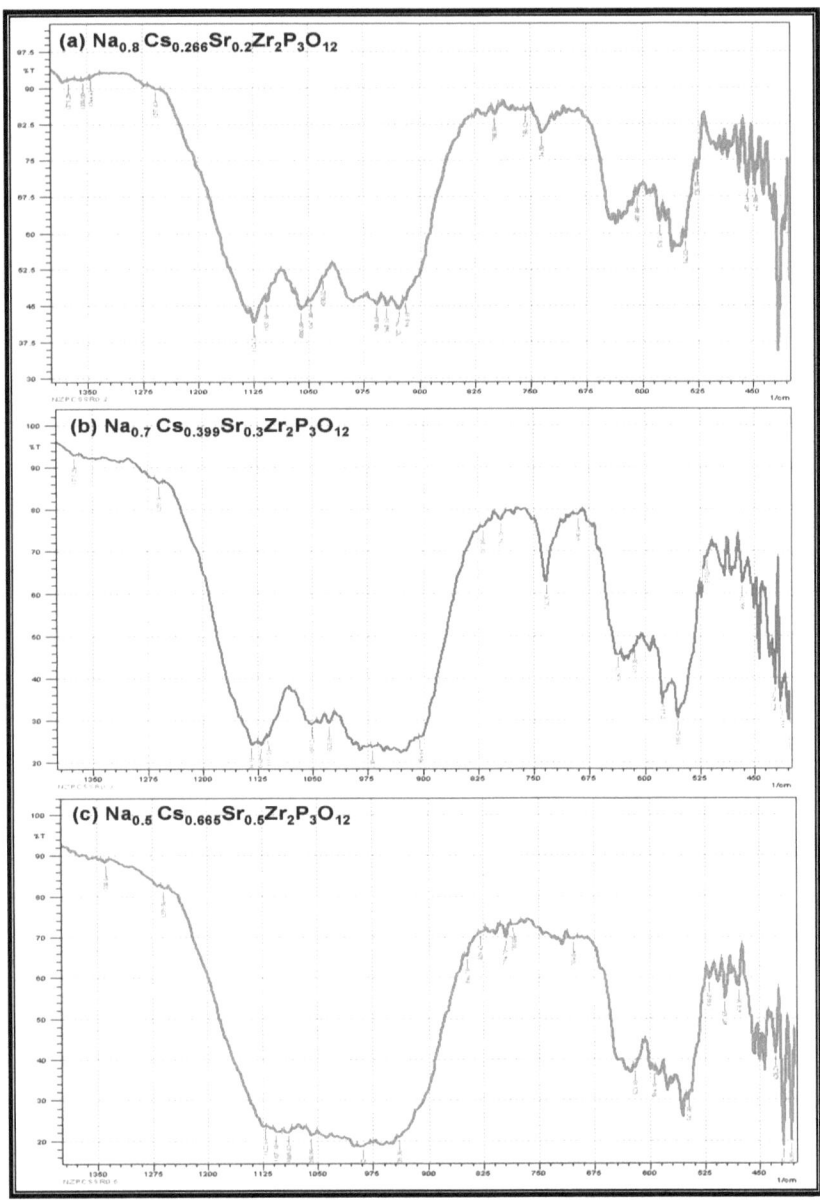

Figure 4.10 Infrared spectra of $Na_{1-x}(Cs_{1.33}Sr_1)_xZr_2P_3O_{12}$ (x = 0.1 to 1.0) (a) x = 0.2, (b) x = 0.3 and (c) x = 0.5 ceramic phases

4.2 Metal Substituted calcium zirconium phosphates:

4.2.1 Molybdenum Substituted calcium zirconium phosphates:

$Ca_{1-2x}Zr_4Mo_{2x}P_{6-2x}O_{24}$ (where x = 0.1-0.5)

Rietveld refinement and crystallographic model of the phases

The powder XRD data showed that monophases of composition $Ca_{1-2x}Zr_4Mo_{2x}P_{6-2x}O_{24}$ (x=0.1-0.3) are isostructural to $CaZr_4(PO_4)_6$. Mo modified CZP crystallizes in the rhombohedral system (space group *R-3*). The conditions for the rhombohedral lattice: (i) -h+k+l=3n (ii) when h=0, l=2n and (iii) when k=0, l=2n have been verified for all reflections between 2θ=10°-90°. The intensity and positions of the diffraction pattern match with the characteristic pattern of parent compound calcium zirconium phosphate, which gives several prominent reflections between 2θ=12°-58° [28]. The General Structure Analysis System (GSAS) program with the EXPGUI graphical user interface was used to carry out crystal structure determination which allows refinement of atomic coordinates, site occupancies and atomic displacement parameters as well as profile parameters (lattice constants, peak shape, peak height, instrument parameters, background).refinement leads to rather good agreement between the experimental and calculated XRD pattern (Fig. 3.31, 3.33 & 3.35) and yields acceptable reliability factors: RF^2, Rp and Rwp . The normal Probability plot for the histogram gives nearly a linear relationship indicating that the *Io* and *Ic* values for the most part are normally distributed (Fig. 3.32, 3.34 & 3.36). The lattice parameters are close to the corresponding values for un-substituted CZP unit cell [29-30]. The cell parameters of the specimens register slight increase in *a* direction (Table 3.11) [31]. Alteration in lattice parameters shows that the network modifies its dimensions to accommodate the cations occupying M_1 site without breaking the bonds. The basic framework of CZP accepts cations of different sizes and oxidation states to form solid solutions but at the same time retaining the overall geometry unchanged [32-33]. The final atomic coordinates and isotropic thermal parameters (Table 3.12), inter-atomic distances (Table 3.13 & 3.14) and bond angles (Table 3.15 & 3.16) are extracted from the crystal information file prepared after final cycle of the refinement. Selected h, k, l values, d-spacing, and intensity data along with observed and calculated structure factors have been listed in table 3.17. The refinement leads to acceptable Zr-O, P-O bond distances. Zr atoms are displaced from the center of the octahedron due to the $Ca^{2+}-Zr^{4+}$ repulsions. Consequently the Zr–O(2) distance, neighboring the calcium Ca(1), is slightly greater than the Zr–O(1) distance, however, average Zr–O distances are smaller than the values calculated from the ionic radii data (2.12 Å). The O–Zr–O angles vary between 79.327° and 178.118°. The angles implying the shortest bonds are superior to those involving the longest ones due to O-O repulsions which are stronger for O(1)-O(1) than for O(1)-O(2).

The P-O distances are close to those found in 'Nasicon' type phosphates. The O-P-O angles vary between 108.494°-119.055°. Fig. 4.11 and 4.12 illustrate the Ball & Stick and diamond view respectively of ceramic phase showing the ZrO_6 inter ribbon distance in the structure of the title phase which is a function of amount and size of alkali cation in the M site of the 3D framework, built from ZrO_6 octahedrons and corner sharing PO_4 tetrahedrons.

SEM and EDAX analysis

The EDAX spectra provide the evidence of Mo in the polycrystalline mono phases while scanning electron micrographs show the typical morphology of the grains. Simultaneously, the particle size was also determined using the Scherer's equation where broadening of peak is expressed as full width at half maxima (FWHM) in the recorded XRD pattern (Table 4.5). The particle size varies between 19.58-152.28 nm. Crystallographic planes and ordered arrangement of atoms is visible in the Scanning electron microscopy image (Fig. 3.37-3.38).

IR Analysis

The presence of orthophosphate anions in the crystal structure was confirmed with the IR spectroscopy. The absorption bands in the range between 1250-1022 cm^{-1} and 650-507cm^{-1} are assigned to stretching and bending vibrations of P-O bonds of the PO_4 tetrahedron respectively. The stretching vibrations occur between 1270-1020cm^{-1} as υ_3 band, the symmetric stretching υ_1 and anti symmetric bending υ_4 vibrations are observed in the regions 990-900cm^{-1} and 640-505 cm^{-1} respectively [34-35] (Table 4.6. and Fig. 4.13).

Table 4.5 Distribution of particle size (nm) along with prominent reflecting planes of $Ca_{1-2x}Zr_4Mo_{2x}P_{6-2x}O_{24}$ (x = 0.1-0.5) ceramic sample

h k l	FWHM (2θ)	Particle size (nm)
$Ca_{0.8}Zr_4Mo_{0.2}P_{5.8}O_{24}$		
1 1 3	0.1506	54.82
2 0 -4	0.1632	54.82
1 1 6	0.1840	44.21
2 1 1	0.1020	97.9
2 1 4	0.0816	114.21
3 0 0	0.1224	76.14
2 0 8	0.0612	152.28
1 1 9	0.1224	72.13
2 2 0	0.1020	97.9
2 1 -8	0.1428	62.3
2 0 -10	0.1632	57.10
4 0 -2	0.1224	80.62
2 1 10	0.0612	152.28
3 1 -7	0.0612	19.58
3 1 8	0.1020	97.9
4 1 0	0.1020	91.37
4 0 -8	0.1004	91.37
4 1 6	0.1428	72.13
4 2 -4	0.1004	91.37
3 2 10	0.0816	114.21
4 3 10	0.0816	124.6

Ca$_{0.4}$Zr$_4$Mo$_{0.6}$P$_{5.4}$O$_{24}$

2 0 -4	0.1840	44.21
1 1 6	0.1840	44.21
2 1 1	0.2007	42.83
2 0 8	0.0816	124.6
2 2 0	0.1338	65.26
1 1 9	0.2342	35.14
1 0 10	0.0836	105.4
3 0 -6	0.1004	85.66
2 1 -8	0.1673	52.17
2 2 6	0.1506	57.10
4 0 -2	0.1338	76.14
3 1 8	0.1171	85.66
4 0 -8	0.0612	152.28
4 1 6	0.0669	152.28
2 1 -14	0.2676	35.14
3 2 10	0.1673	57.10
5 2 0	0.1338	65.26
4 3 -8	0.2342	40.31

Zr$_4$MoP$_5$O$_{24}$

1 0 -2	0.2040	44.21
1 1 0	0.1673	50.76
1 1 3	0.1673	48.95
1 1 6	0.1171	72.13
2 1 1	0.1004	85.66
2 1 4	0.1632	59.59

2 0 8	0.1224	80.62
1 1 9	0.1632	62.3
2 2 0	0.1020	97.9
3 0 -6	0.0816	114.21
2 1 -8	0.0836	114.21
3 1 -4	0.1224	80.62
4 0 -2	0.0816	114.21
2 1 10	0.0612	152.2
3 2 4	0.2342	37.04
4 1 0	0.1566	52.71
1 0 -14	0.2448	37.04
4 1 -6	0.1632	54.82
2 0 14	0.1020	97.9
3 3 0	0.2007	42.83
5 1 4	0.0816	114.21
6 0 0	0.0816	124.6
3 2 -14	0.1224	80.62

Table 4.6 Assignment (cm^{-1}) of IR bands for $Ca_{1-2x}Zr_4Mo_{2x}P_{6-2x}O_{24}$ (x = 0.1-0.5) ceramic samples

Compound	υ_3 υ_{as} (P-O)	υ_1 υ_a (P-O)	υ_4 δ (P-O)	υ_2 (P-O)
$Ca_{0.8}Zr_4Mo_{0.2}P_{5.8}O_{24}$	1043.52 1074.39 1087.89 1101.39 1230.63	945.15 956.72	553.59 636.53	406.99 428.21 466.79
	1033.88	900.79	528.51	401.21

Ca$_{0.4}$Zr$_4$Mo$_{0.6}$P$_{5.4}$O$_{24}$	1070.53	929.72	553.59	416.64
	1103.32	970.23	572.88	434
	1170.53	981.8	636.53	443.63
	1224.84			462.93
Zr$_4$MoP$_5$O$_{24}$	1049.31	920.08	522.73	403.14
	1064.74	943.22	536.23	408.92
	1076.32	964.44	574.81	426.28
	1097.53	987.59		
	1184.33			
	1205.55			

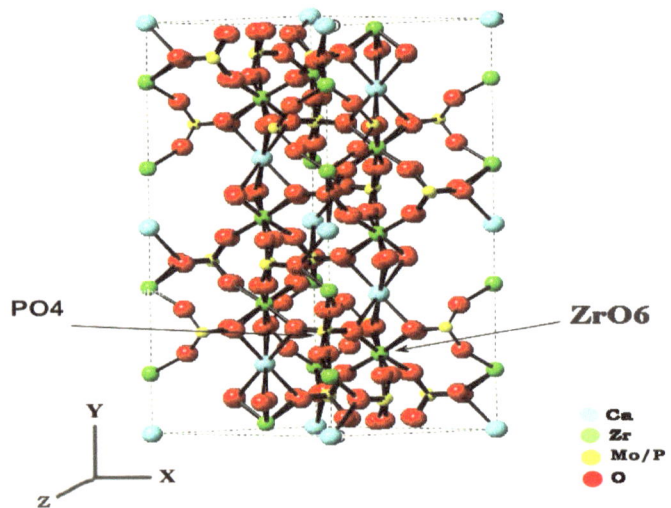

Fig.4.11 Ball and stick representation of Ca$_{0.8}$Zr$_4$Mo$_{0.2}$P$_{5.8}$O$_{24}$ ceramic sample

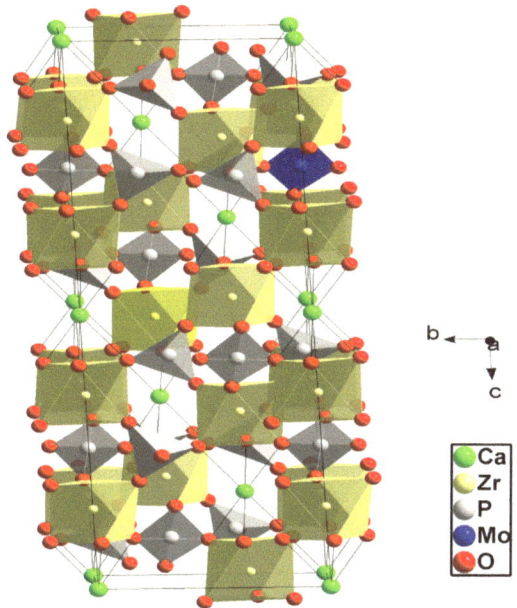

Fig.4.12 DIAMOND view of crystal structure of $Ca_{0.8}Zr_4Mo_{0.2}P_{5.8}O_{24}$ ceramic phase

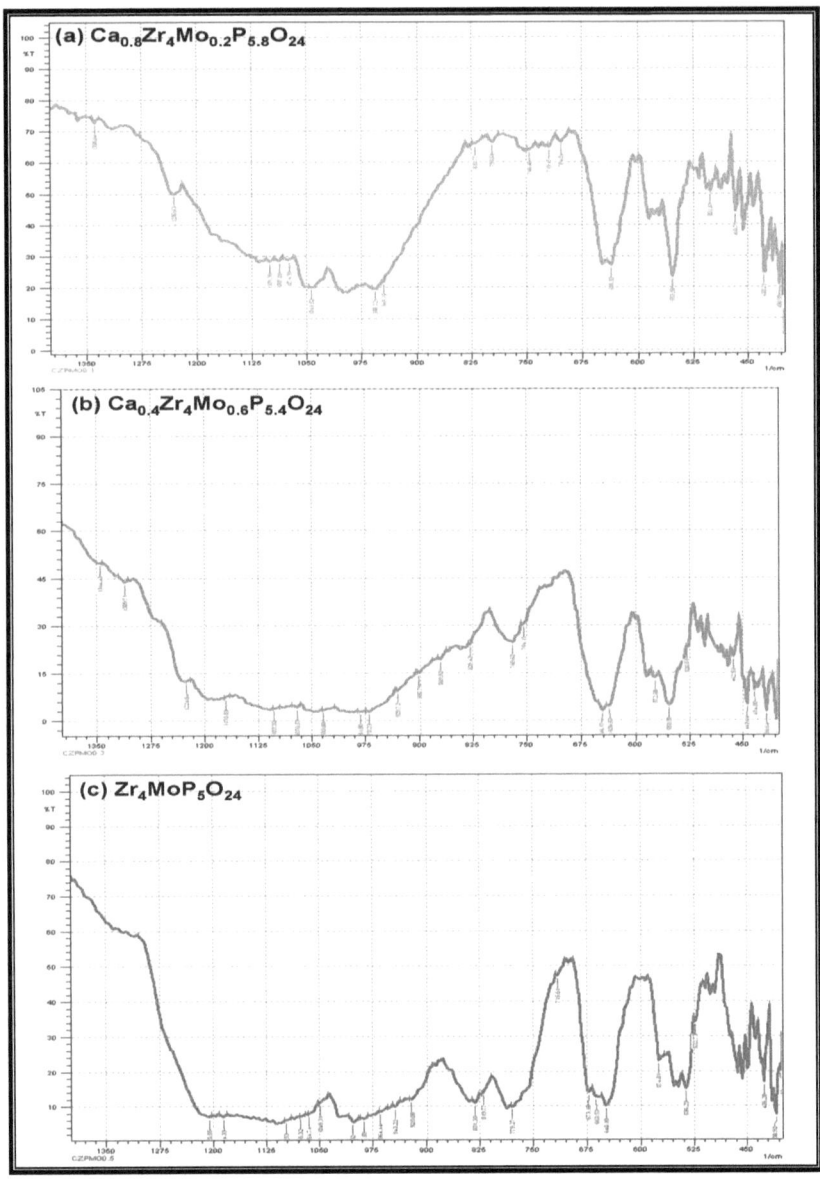

Fig 4.13 Infrared spectra of $Ca_{1-2x}Zr_4Mo_{2x}P_{6-2x}O_{24}$ (a) x = 0.2, (b) x = 0.3, (c) x = 0.5 Ceramic samples

4.3 Perovskites

4.3.1 Calcium Samarium Titanate: $Ca_{1-x}Sm_xTiO_3$ (where x = 0.1-0.5)

Crystallographic data

The X-ray diffraction patterns (Fig. 3.39) match in intensity and positions with those of parent perovskite structure of $CaTiO_3$ which indicating that the matrix can be modified by partial substitution of calcium by samarium to yield a single phase polycrystalline solid solution [36]. The Rietveld refinement of crystal data of phase pure Calcium samarium titanate was performed by the least square method using GSAS software. Rietveld plot of this compound matches with the Bragg positions for the proposed structure model (Fig. 3.40, 3.42, 3.44 & 3.46). The pattern was indexed for orthorhombic system constrained by operators of space group Pbnm (#62). The perovskite stoichiometry is expressed as ABO_3 where the size of the 'A' cation (divalent) is substantially large to form a close packed array with oxygen anion while the 'B' (tetravalent) ions must fit within the octahedral holes of the A–O close packed array. The ideal perovskite structure is cubic with a = 3.8 Å. The structure consists of B site cations that are octahedrally coordinated by atoms of oxygen. Larger A site cations are surrounded by twelve atoms of oxygen in cubo – octahedral coordination within this framework. The reduction in symmetry from cubic to orthorhombic results from the presence in the A site of cations that are smaller than required to maintain the ideal geometry of this site.

The reflection conditions for this space group were verified from the international table for X-ray crystallography and further checked by presence of *021* and absence of *201* reflections, which are not allowed [37-38]. Other systematically absent reflections in support of chosen space group were also verified. In addition to this both odd-odd-odd and odd-odd-even reflections are present in the title phase indicating that the oxygen octahedra are tilted both in phase as well as anti phase [39].

The normal probability plot for the histogram gives nearly a straight line indicating that the *Io* and *Ic* values are for the most part normally distributed (Fig. 3.41, 3.43, 3.45 & 3.47) with slop of 1.352, 1.795, 1.531 &1.551 for

$Ca_{1-x}Sm_xTiO_3$ (x=0.1 – 0.5) ceramic phases. The slope is characteristic of the completeness of the refinement, as more parameters are refined and a better fit is obtained this value will approach 1.0. The structural solution was derived from the crystal information file prepared after final cycle of refinement by Rietveld analysis. The accuracy of the fit is indicated by the agreement in the expected and calculated R factors and goodness of fit and yields acceptable reliability factors: Rp, Rwp and RF^2 (Table 3.18).The atomic coordinates and isotropic thermal parameters are given in Table 3.19. The major interatomic distances are summarized in Table 3.20 while Table 3.21 lists the bond angle's data. The inter-atomic distances between Ti (5) and the apex oxygen atoms of the octahedron have been found to be 1.92311 Å whereas the two sets of planar oxygen bond distances Ti(5)-O(1) are 1.38192 and 2.46815 Å respectively. The six fold coordination is maintained for the B site at centre of the perfect octahedra as indicated by the Ti-O bond lengths. The framework structure of ceramic is constructed of corner sharing TiO_6 octahedra and with Sm^{3+} ions placed in twelve co-ordinate interstices (Fig. 4.30). For given coordination number the calculated M-O distances are close to their standard values of metal oxygen distances in corresponding metal oxides. The distance and angle data indicate slight distortion of the TiO_6 framework. The observed orthorhombic distortion from the ideal cubic symmetry of perovskites may be described as a result of tilting of the oxygen octahedra surrounding the Ti atoms. The B site of the perovskite maintains six co-ordinations with the shortest Ti(5)-O (1) bond of 1.38192 Å and the longest one Ti(5)-O(2) of 2.46815 Å. The distortion of TiO_6 reduces the coordination number of Ca/Sm atoms to 8 resulting into eight acentric bonds in contrast to twelve for ideal cubic perovskite. The geometry of Ca/Sm coordination sphere defined by the tilted octahedra is, therefore, a distorted cuboctahedron in relation to ideal A site polyhedron in cubic perovskite, which has four triangular and six square faces [40]. Table 3.22 summarizes the polyhedral distortions 'Δ' [41] for CaO_8 and TiO_6 polyhedrons.

The scanning electron micrographs show the typical morphology of the grains (Fig. 3.48). The EDX analysis shows consistency in observed and theoretical atomic and weight ratios with respect to Ca, Sm and Ti .Within the permissible statistical limits. The EDX spectra show the presence of Samarium in the polycrystalline phases (Fig. 3.49).

X-ray data for the samples were used to estimate the particle size using Scherer's formula. The *h, k, l* values corresponding to prominent reflections and particle size for crystals are shown in Table 3.23.The particle size calculation (Table 4.7) has been done for most of the low angle and high intensity reflections. The particle size distribution along the reflection planes ranges from 11.23 - 105.43 nm for $Ca_{1-x}Sm_xTiO_3$ (x = 0.1 – 0.5) ceramic powders.

Electrical properties

The dielectric constant (ε') and loss (tanδ) have been investigated over a wide frequency (100 Hz – 25 MHz) and temperature (25°C - 300°C) range. The effect of substituent (Sm) on the dielectric behavior and Curie temperature was studied. The results of my investigations are discussed below:

Dependence of ε' and tanδ on frequency

Fig. 4.14-4.21 show the frequency dependence of dielectric constant and loss at different temperatures for the Calcium Titanate modified by the substitution of Sm. Dielectric constant & loss factor (ε' and tanδ) shows decreasing trend with increase in frequency. The dielectric constant of $Ca_{1-x}Sm_xTiO_3$ (x= 0.1 to 0.5) at lower frequencies is higher than the corresponding values in higher frequency region. This is the normal behavior of ferroelectric materials [42].

It was observed that the value of dielectric constant of each specimen at higher frequency gets markedly dropped. This phenomenon can be explained in term of interfacial polarization. This built up of charges at the grain-grain boundary interface is responsible for large polarization, therefore high dielectric constant at low frequency.

The fall in dielectric constant arises from the fact that polarization does not occur instantaneously with the application of the electric field because of inertia. The delay in response towards the impressed alternating electric field leads to loss and decline in dielectric constant. At low frequencies, all types of polarization such as interfacial, atomic, dipolar, ionic and electronic contribute. As frequency is increased, those with large relaxation times cease to respond and hence the decrease in dielectric constant. While studying the behavior of dielectric constant with frequency, it is important to note that at lower frequencies, the contribution from the space charge polarization is maximum, which reduces slowly with the increase of frequency. The space charge arises from the charge accumulation at grain boundaries and at the electrode interface, mainly due to vacancies of calcium and oxygen. Since with the increase of frequency, these dipoles due to space charge do not respond at higher frequencies resulting in decrease in the dielectric constant. The same type of frequency dependent dielectric behavior is found in many ferroelectric ceramics [43-44]. The behavior of tanδ also follows the same reasoning [45-47]. The dielectric relaxation phenomenon in ferroelectric materials reflects the delay (time dependence) in the frequency response of a group of dipoles when submitted to an external applied field. When an alternating voltage is applied to a sample, the dipoles responsible for the polarization are no longer able to follow the oscillations of the electric field at certain frequencies. The field reversal and the dipole reorientation become out-of-phase giving rise to a dissipation of energy. Over a wide frequency range, different types of polarization cause several dispersion regions and the critical frequency, characteristic of each contributing mechanism, depends on the nature of the dipoles.

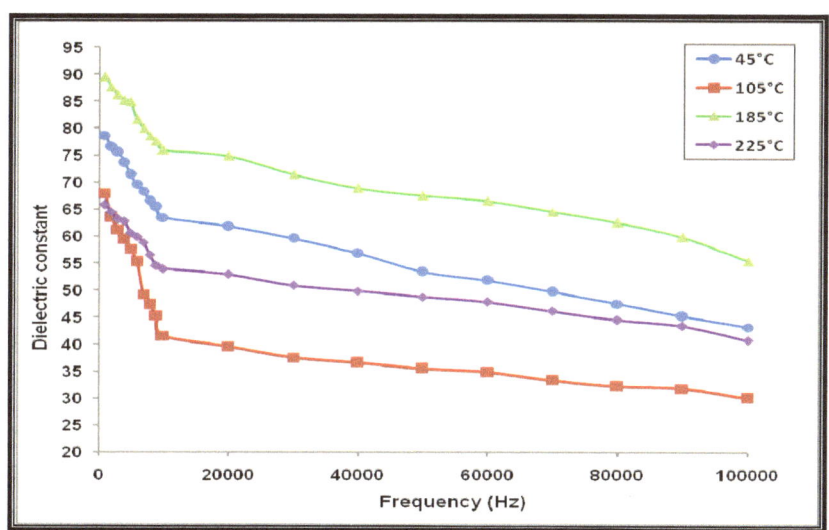

Fig 4.14 Variation of ε' with frequency at different temperature for $Ca_{0.9}Sm_{0.1}TiO_3$

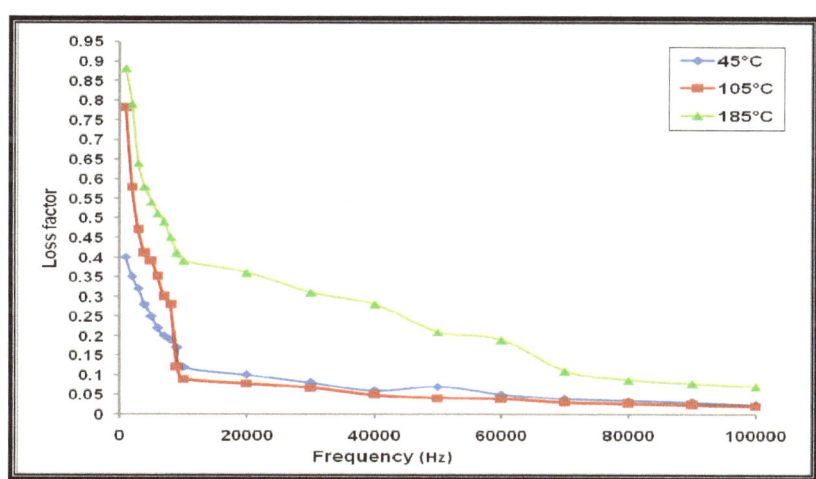

Fig 4.15 Variation of tanδ with frequency at different temperature for $Ca_{0.9}Sm_{0.1}TiO_3$

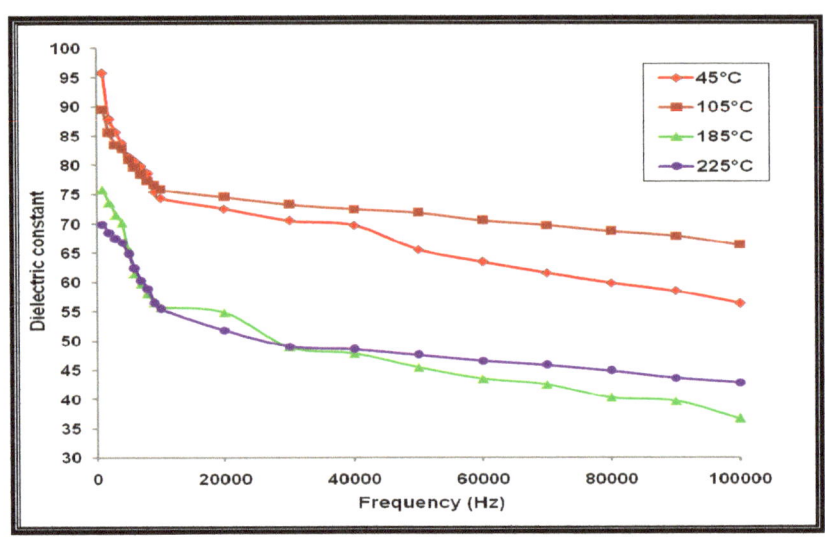

Fig 4.16 Variation of ε' with frequency at different temperature for $Ca_{0.8}Sm_{0.2}TiO_3$

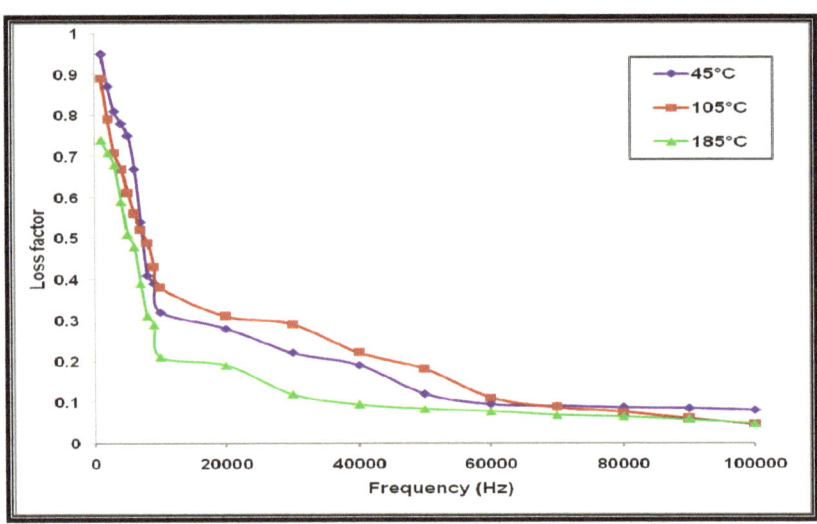

Fig 4.17 Variation of tanδ with frequency at different temperature for $Ca_{0.8}Sm_{0.2}TiO_3$

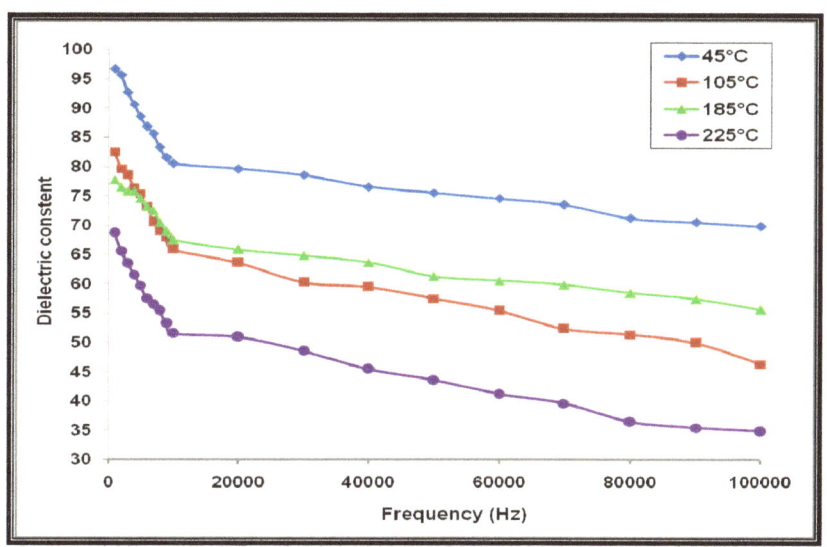

Fig 4.18 Variation of ε' with frequency at different temperature for $Ca_{0.7}Sm_{0.3}TiO_3$

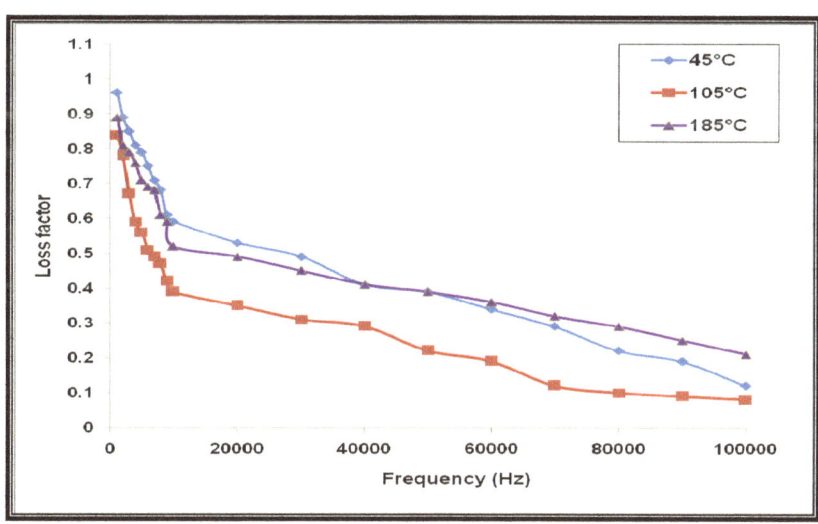

Fig 4.19 Variation of tanδ with frequency at different temperature for $Ca_{0.7}Sm_{0.3}TiO_3$

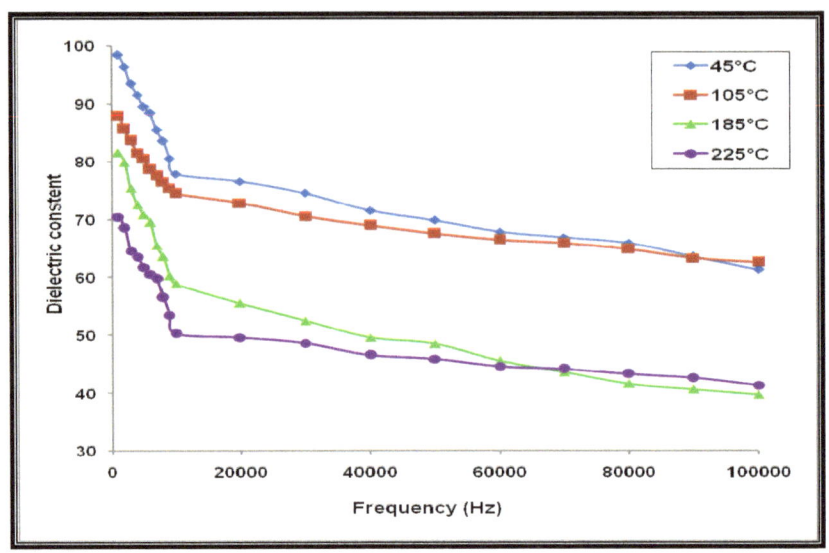

Fig. 4.20 Variation of ε' with frequency at different temperature for $Ca_{0.6}Sm_{0.4}TiO_3$

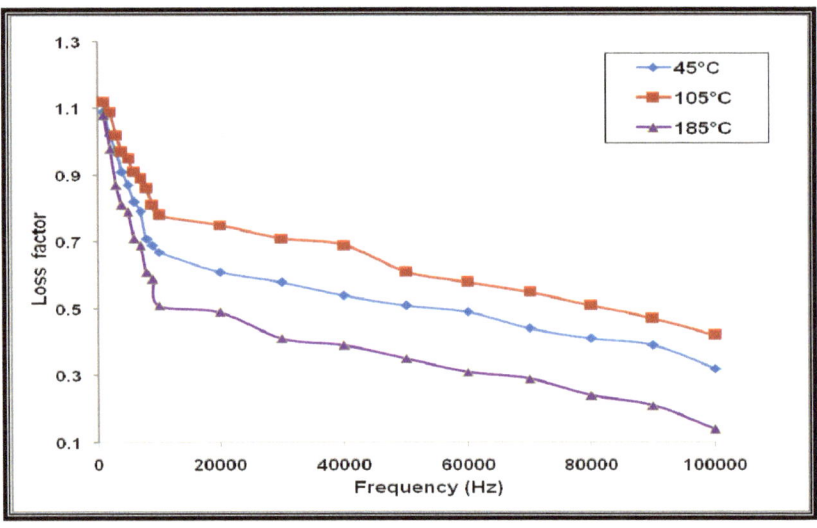

Fig. 4.21 Variation of tanδ with frequency at different temperature for $Ca_{0.6}Sm_{0.4}TiO_3$

Dependence of ε' and tanδ on temperature

The dependence of dielectric constant and loss factor at different temperature was shown in figure (4.22-4.29). As typical of normal ferroelectrics, dielectric constant increases gradually with increment in the temperature due to interfacial polarization becoming more dominant as compared to the dipolar polarization and passes through a maximum (Curie temperature, T_c) and then decreases due to the phase transition from ferroelectric to the paraelectric phase.

The temperature dependent dielectric loss (tanδ) slowly increases with increase of temperature at all frequencies and temperatures up to 300°C. The variation of tanδ at low temperature is smaller compare to that at higher temperature. At higher temperature conductivity begin to dominate resulting in an increase of tanδ, and hence at higher temperature tanδ is typically associated with the loss by conduction [48-52]. In the dielectric materials defects, space charge formation, lattice distortions etc in the boundaries produce an absorption current resulting in dielectric loss. The main source of dielectric loss in ceramic materials is photon absorption, which is associated with structural disorder such as creation of oxygen deficiency in the material due to doping [53-56]. The tangent loss depends on the dielectric constant as $\tan \delta = \varepsilon''/\varepsilon'$, so it varies proportionally with ε'' In the low frequency range, ε'' is dominated by the influence of ion conductivity. The change of ε'' in the lower frequency range is mainly caused by dipolar relaxation due to atomic and electronic polarizations [57]. Thus, the variation of ε and tanδ indicates the presence of relaxor properties in the material. The relaxor behavior is induced due to many factors such as microscopic composition fluctuation, the merging of micro-polar regions into macro-polar regions or coupling of order parameters and local disorder mode through the local strain [58-61].

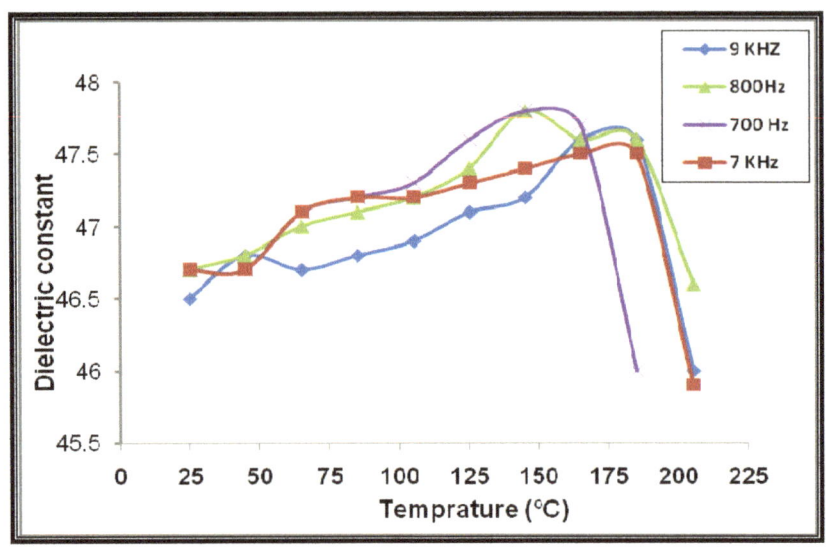

Fig. 4.22 Variation of ε' with temperature at different frequency for $Ca_{0.9}Sm_{0.1}TiO_3$

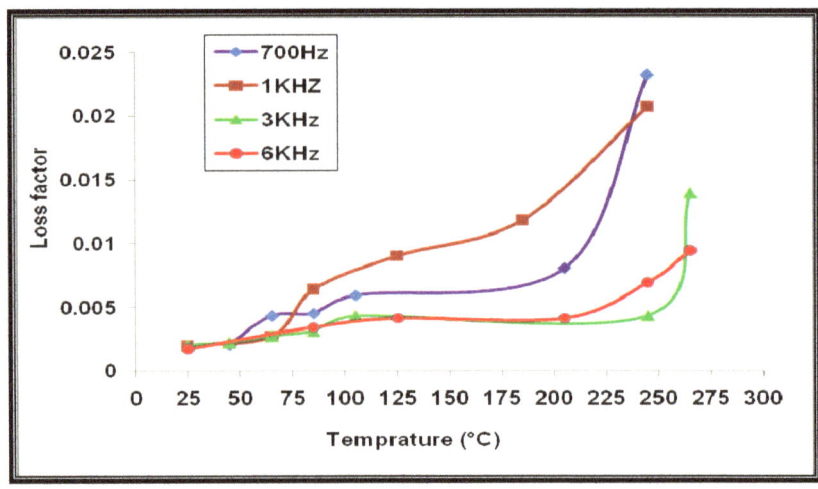

Fig. 4.23 Variation of tanδ with temperature at different frequency for $Ca_{0.9}Sm_{0.1}TiO_3$

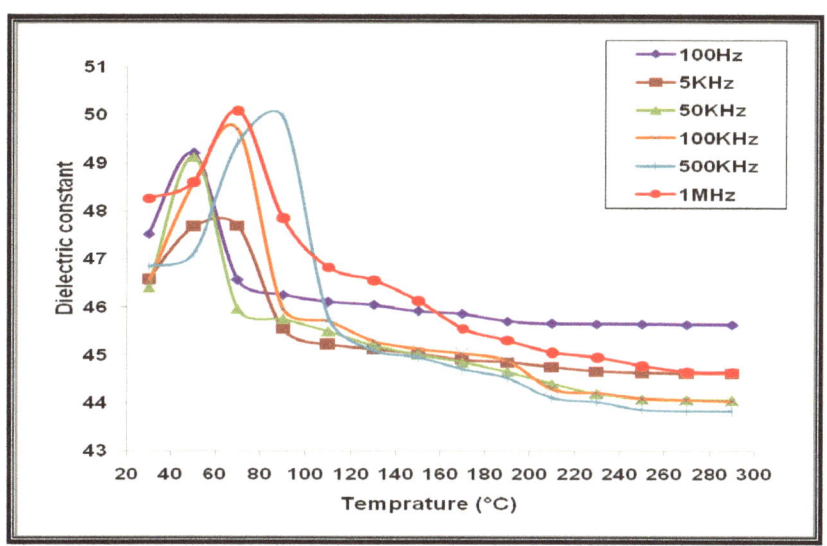

Fig. 4.24 Variation of ε' with temperature at different frequency for $Ca_{0.8}Sm_{0.2}TiO_3$

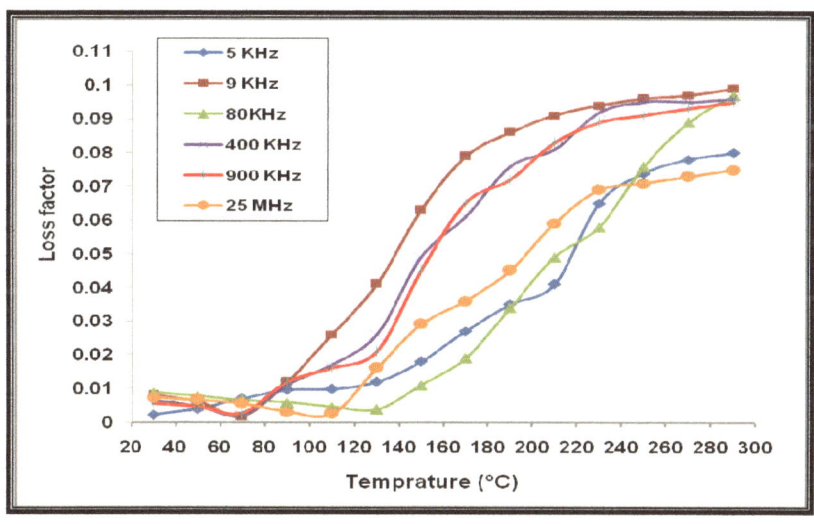

Fig. 4.25 Variation of tanδ with temperature at different frequency for $Ca_{0.8}Sm_{0.2}TiO_3$

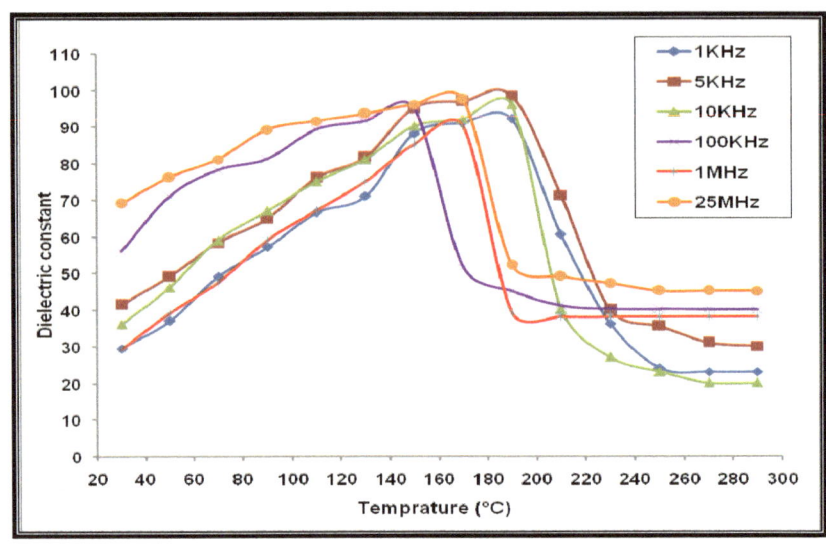

Fig. 4.26 Variation of ε' with temperature at different frequency for $Ca_{0.7}Sm_{0.3}TiO_3$

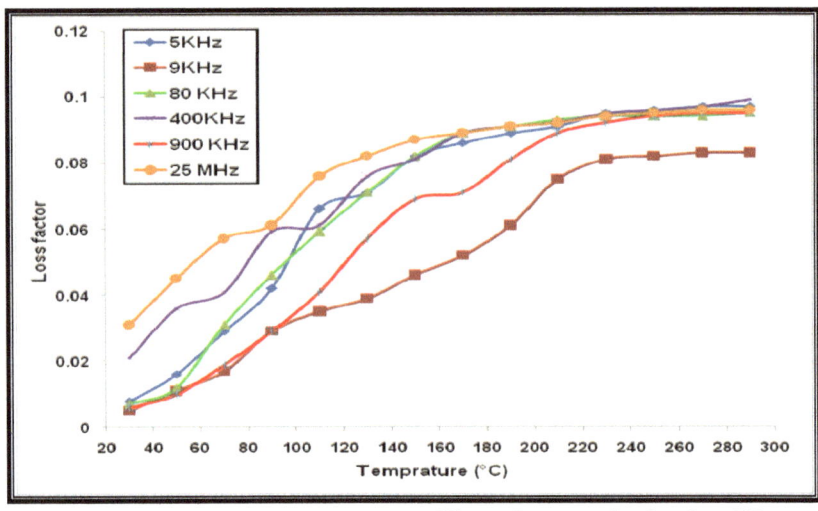

Fig. 4.27 Variation of tanδ with temperature at different frequency for $Ca_{0.7}Sm_{0.3}TiO_3$

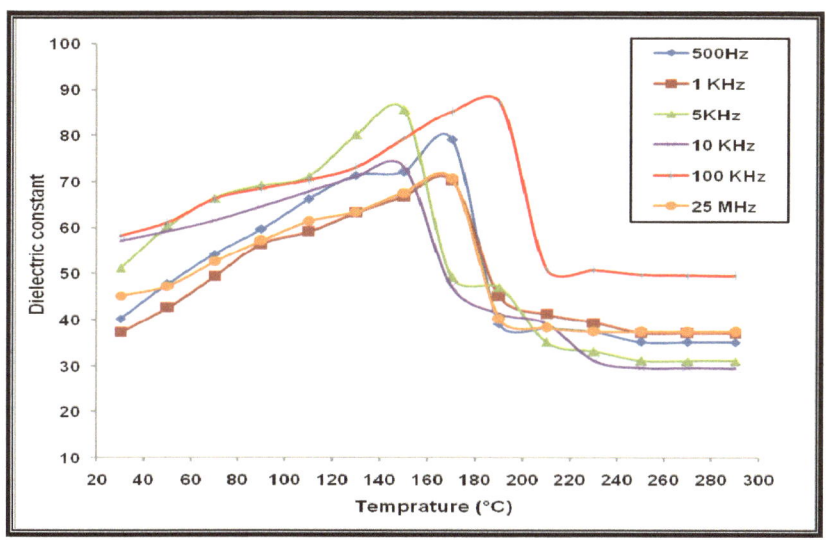

Fig. 4.28 Variation of ε' with temperature at different frequency for $Ca_{0.6}Sm_{0.4}TiO_3$

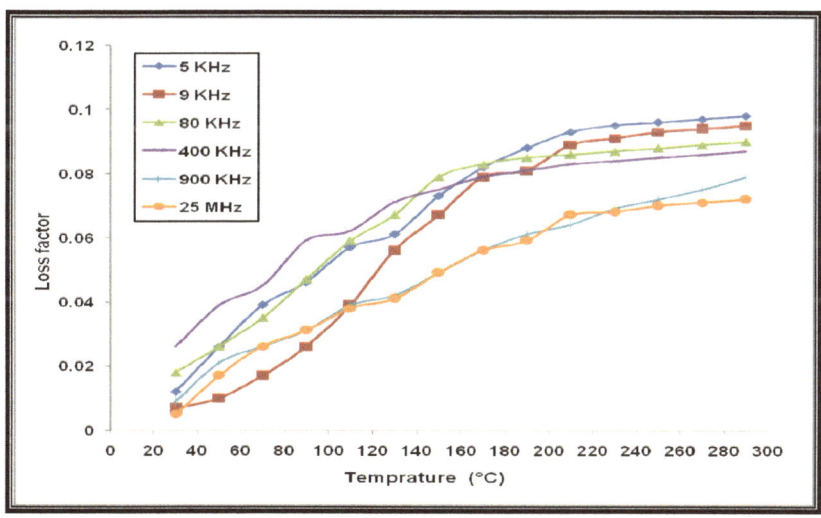

Fig. 4.29 Variation of tanδ with temperature at different frequency for $Ca_{0.6}Sm_{0.4}TiO_3$

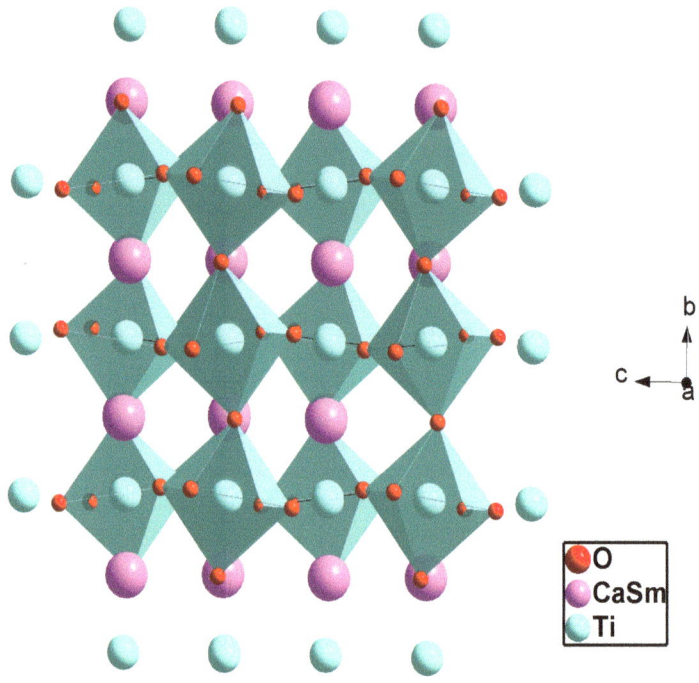

Fig. 4.30 DIAMOND view of crystal structure of $Ca_{0.7}Sm_{0.3}TiO_3$

Table 4.7 Distribution of particle size (nm) along with prominent reflecting planes of $Ca_{1-x}Sm_xTiO_3$ (x = 0.1 - 0.5) ceramic sample

h k l	FWHM (2θ)	Particle size (nm)
$Ca_{0.90}Sm_{0.10}TiO_3$		
1 2 1	0.1338	62.3
2 1 0	0.1004	91.37
2 2 1	0.2007	48.95
2 3 2	0.4015	21.41
0 0 4	0.2007	44.21
4 1 0	0.2007	27.41

3 2 3	0.1004	91.37
2 0 4	0.1338	62.3
4 3 2	0.2448	47.26
2 2 0	0.1338	62.3
1 6 1	0.1338	65.26
1 1 1	0.1338	65.26
0 0 2	0.1004	105.4
0 2 2	0.1673	62.3
$Ca_{0.8}Sm_{0.2}TiO_3$		
1 2 1	0.0836	105.4
2 1 0	0.2007	42.83
4 4 0	0.4015	21.08
1 0 1	0.5353	15.93
0 3 1	0.5353	16.31
2 1 1	0.6691	14.29
0 4 2	0.1338	68.53
3 2 1	0.2007	45.68
2 4 2	0.1004	105.43
3 2 3	0.6528	16.31
$Ca_{0.7}Sm_{0.3}TiO_3$		
2 0 0	0.2448	38.07
1 2 1	0.1171	72.13
2 0 1	0.3264	29.16
0 3 1	0.2007	41.53
0 2 2	0.1338	62.30
2 3 2	0.3264	27.41

1 6 1	0.1338	65.26
3 5 1	0.2007	41.53
3 3 3	0.3346	25.86
0 4 4	0.1673	50.76
4 1 2	0.2448	37.04
3 1 1	0.4080	25.38
2 0 4	0.2007	41.53
$Ca_{0.6}Sm_{0.4}TiO_3$		
1 0 1	0.2007	40.31
1 1 1	0.2007	44.21
1 2 1	0.0836	105.4
2 1 1	0.2342	39.16
0 2 2	0.2007	44.21
1 3 1	0.4896	22.10
1 2 2	0.9368	11.23
0 4 2	0.2676	35.14
2 1 3	0.4015	20.45
3 3 1	0.2676	35.14
2 4 2	0.1338	62.30
4 0 0	0.1338	65.26
2 0 4	0.2007	41.53
3 2 3	0.1004	56.66
4 0 2	0.2007	44.21
3 3 3	0.1004	97.90
4 4 0	0.1171	72.13
0 4 4	0.4015	20.45

References

[1] JCPDS Powder diffraction data file no.71-0959 compiled by International Center for Diffraction Data, USA, **(2000)**.

[2] Larson AC, Von Dreele RB, General structure analysis system technical manual. LANSCE, MS-H805, Los Almos National Laboratory LAUR **(2000)**: 86-748.

[3] Toby, B. H., EXPGUI, a graphical user interface for GSAS. Journal of Applied Crystallography, 34 **(2001)**: 210.

[4] Rietveld H M J Appl Cryst 2. **(1969)**: 65.

[5] I. Yu. Pinus, A. R. Shaikhlislamova, I. A. Stenina, N. A. Zhuravlev and A. B. Yaroslavtsev. Inorganic Materials, 45[12], **(2009)**: 1370.

[6] Kojitani H, Kido M, Akaogi M Phys Chem Miner 32. **(2005)**: 290.

[7] Verissimo C, Garrido FMS, Alves OL, Calle P, Juarez AM,Iglesias JE, Rojo JM Solid State Ionics 100, **(1997)**: 127.

[8] Govindan Kutty KV, Asuvathraman R, Sridhran R J Mater Sci 33. **(1998)**: 4007.

[9] Shannon R D Acta Crystallogr A 32. **(1976)**: 751.

[10] A. I. Pylinina and I. I. Mikhalenko Russian Journal of Physical Chemistry A, 85[12], **(2011)**: 2109.

[11] A. I. Pylinina, I. I. Mikhalenko, M. M. Ermilova, et al., Zh. Fiz. Khim. Russian Journal of Physical Chemistry A, 84[12], (**2010**): 465.

[12] N. Anantharamulu et. Al. J Mater Sci 46, **(2011)**: 2821.

[13] Nunotani N, Sawada M, Tamura S, Imanaka N Bull Chem Soc Jpn 83, **(2010)**: 415.

[14] Navulla A Solid State Ion 181 **(2010)**: 659.

[15] H. M. Roger, P. L. Ruslan, J. Solid State Chem., 177, **(2004)**: 4420.

[16] M. A. Subramanian, J.C. Calabrese, Mater. Res. Bull., 28, **(1993)**: 523.

[17] E.T. Keve, A.C. Skapski, J. Solid State Chem., 8, **(1973)**:159.

[18] E.S. Kim, K.H. Yoon, J. *Eur. Ceram. Soc.* 23, **(2003)**: 2397.

[19] N.Bhatt, Rajiv Vaidya, S. G. Patel, A. R. Jani, *Bull. Mater. Sci.* 27 **(2004)**: 23.

[20] Shrivastava OP, Chourasia R, Kumar N Ann Nucl Energy 35–36: **(2008)**: 1147.

[21] V. I. Pet'kov, E. A. Asabina, A. V. Markin and N. N. Smirnova Journal of Thermal Analysis and Calorimetry, 91, **(2008)**: 155.

[22] E. Yu. Borovikova, V. S. Kurazhkovskaya D. M. Bykov, and A. I. Orlova. Journal of Structural Chemistry. 51[1]. **(2010)**: 40.

[23] V.S. Kurazhkovskaya, D.M. Bykov, E.Yu. Borovikova, N.Yu. Boldyrev, L. Mikhalitsyn, A.I. Orlova. Vibrational Spectroscopy, 52[2], **(2010)**:137.

[24] M. V. Sukhanov, V. I. Pet'kov, and D. V. Firsov Inorganic Materials, 47[6] **(2011)**: 674.

[25] M. V. Sukhanov, V. I. Pet'kov and D. V. Firsov, Russian Journal of Inorganic Chemistry, 56[9], **(2011)**: 1351.

[26] V. I. Pet'kov et. al. Inorganic Materials, 47[2], **(2011)**: 178.

[27] Pet'kov, V.I., Asabina, E.A., Markin, A.V., and Smirnova, N.N., *J. Chem. Eng. Data,* 55[2], **(2010)**: 856-863.

[28] International Tables for X-ray crystallography: Natl. Bur. Stand (U.S.) Monogr. 25, 20, 36 **(1984)** Card No.33-0321.

[29] V. I. Pet'kov, A. V. Markin, I. A. Shchelokov, N. N. Smirnova and M. V. Sukhanov, Russian Journal of Physical Chemistry A, 84[4], **(2010)**: 541.

[30] Ute Ploska and Georg Berger, *Biomaterials* 18 **(1997)**: 1671.

[31] Chourasia Rashmi and O.P. Shrivastava, Solid State Sciences 13 **(2011)**: 444.

[32] J. Alamo, and R.Roy, J. Am. Ceram. Soc., 67, **(1984)**: 78.

[33] G. E. Lenain, H. A. McKinstry, J Alamo and D. K. Agrawal, J. Mat. Sci., 22, **(1987)**: 17.

[34] M. Barj, H. Perthuis, Ph.Colomban Solid State Ionics 11. **(1983)**: 157.

[35] A. Mbandza, E. Bordes, P. Courtine Mater. Res. Bull. 20. **(1985)**: 251.

[36] D. M. Thomas and L. Andrews, J. Mol. Spectrosc. 50. **(1974)**: 220.

[37] X. Liu, R. C. Liebermann, *Phy. Chem. Miner.* 20. **(1993)**:171.

[38] International Tables for X-ray crystallography: Kynoch Press Birmingham 4, 1974 Card No.82-0232.

[39] R. J. Bouchard, J.F. Weither, *J. Solid State Chem.* 4. **(1972)**:80.

[40] A. M. Glazer, *Acta Cryst.* A31. **(1976)**:756.

[41] N. W. Thomas, *Acta.Crys.* B47 **(1991)**: 180.

[42] Eung Soo Kim, Byung Sam Chun, Dong Ho Kang, *J. Europ. Ceram. Soc.* 27 **(2007)**: 3005.

[43] J. Mal and R.N.P. Choudhary, Phase Transitions, 62 **(1997)**: 9.

[44] N.K. Mishra, R. Sati and R.N.P. Chaoudhary, Mater. Lett, 24 **(1995)**: 313.

[45] Vishal Singh, K.K. Bamzai, Shivani Suri, Nidhi. Ceramics International 37 **(2011)**: 2655.

[46] Mohit Sharma, Himani Sharma, K.K. Raina, Journal of Physics and Chemistry of Solids 69 **(2008)**: 2584.

[47] Sarabjit Singh, Chandra Prakash, K.K. Raina Journal of Alloys and Compounds 492 **(2010)**: 717.

[48] A. G. Belous, *J. Eur. Ceram. Soc.*21. **(2001)**: 2717.

[49] Kingery W D, Bowen H K & Uhlmann D R, Introduction to ceramics, 2nd ed (John wiley & sons), **(1976)**: 937.

[50] Kuharuangrong S, J.Mater Sci, 36(7). **(2001)**: 1727.

[51] Robert C. Pullar, Yong Zhang, Lifeng Chen, Shoufeng Yang and Julian R. G. Evans, et al. Journal of Electro ceramics, 22[1-3], **(2009)**: 245.

[52] Jinliang He, Fengchao Luo, Jun Hu and Yuanhua Lin. Science china, Technological Sciences, 54[9], **(2011)**: 2506.

[53] Tickoo R,Tondon,R P,Mehara N C & P N, Mater sci & Eng B,94. **(2002)**: 1.

[54] Dalveer Kaur, Narang Sukhleen Bindra, K. Singh, *Ceramics International.* 33.**(2007)**: 249.

[55] Robert C. Pullar, Yong Zhang, Lifeng Chen, Shoufeng Yang and Julian R.G. Evans, et al. Journal of Electroceramics, 22[1-3], **(2009)**: 245.

[56] Dong Zhang, Chunli Zhang, Pin Zhou, Journal of Hazardous Materials 186 **(2011)**: 971.

[57] L.F. Chen, C.K. Ong, C.P. Neo, V.V. Varadan, V.K. Varadan, in: *Microwave Electronics*: Measurement and Materials Characterization, John Wiley & Sons, Ltd., **(2004)**.

[58] Tanmoy Maiti, E. Alberta, R. Guo, A.S. Bhalla, Mater. Letters. 60. **(2006)**: 3861.

[59] Masatomo Yashima, Roushown Ali. Solid State Ionics 180 **(2009)**: 120..

[60] L. E. Cross, *Ferroelectrics* 76. **(1987)**: 241.

[61] F. Sánchez-De Jesús, C.A. Cortés-Escobedo, A.M. Bolarín-Miró, A. Lira-Hernández, G. Torres-Villaseñor, Ceramics International 38 **(2012)**: 2139-2144.

Conclusion

Syntheses of all the three groups of materials (NZP, CZP and Perovskite) have been successfully carried out by solid state ceramic route resulting in the formation of single phase high density material in all the cases. Preliminary characterization was carried out by X-ray diffraction followed by complete crystal characterization by Rietveld method. Their crystal chemistry has been investigated using software's like CRYSFIRE, Check cell, Winploter etc. Finally, structures have been refined to a satisfactory convergence using GSAS software. The structure model in each case has been given with the help of graphics software's like PLATON, Crystal maker, Diamond and ORTEP etc.

Particle size calculation using Scherrer's formula suggest that crystallite size along prominent reflections belongs to nano range in most of the synthetic phases. On the basis of results emerged after Rietveld refinement of each group of materials, following conclusions can be drawn:

Substituted sodium zirconium phosphates (NZP)

Refinement of powder X-ray diffraction data of NZP related phases show that the solid solutions of substituted sodium zirconium phosphates (NZP) crystallize in the rhombohedral system with space group R-3c. Crystal data and structural parameters of each phase have been refined to a satisfactory convergence with reasonable values of Rietveld parameters (R_p, R_{wp} and RF^2). The calculated values of P–O and Zr–O bond lengths and O–M–O (M= Na, Zr and P) bond angles are close to their corresponding expected values.

Crystal structure of the system: $Na_{1-x}Sr_{x/2}Zr_2P_3O_{12}$ (x=0.1-1.0) has been refined by least square method. The presence of orthophosphate anions was confirmed by their characteristic IR bands in the region of 1280 cm^{-1} to 1020 cm^{-1} are assigned to the stretching asymmetric vibrations v_3, and bands in the region of 980 cm^{-1} to 915 cm^{-1} correspond to the stretching symmetric vibrations v_1 of the $(PO_4)^{3-}$ ions. Bands in the 670-400 cm^{-1} assigned to the bending vibrations v_4 and v_2. The EDX analysis confirms the fixation of substituted ions in the ceramic matrix. Principally phase pure strontium containing NZP formulations can be prepared with simulated strontium loadings up to ~1.98 mol% (6.07 wt%), beyond these limits traces of minor secondary phase of strontium zirconium phosphate starts appearing along with the solid solution. Analytical evidence has been found to conclude that the Strontium is crystallochemically fixed in the ceramic matrix.

The structural analogue of sodium zirconium phosphate $NaZr_2(PO_4)_3$ is a potential material for immobilization of cesium and strontium from radio active waste along with other radio nuclides. Crystal structure of the system: $Na_{l-x}(Cs_{1.33}Sr_1)_xZr_2P_3O_{12}$ (x=0.1-1.0) has been refined by least square method. The existence of Cesium, strontium and several other radionuclide containing NZP-type structures was assessed on the basis of X-ray, SEM/EDX and IR spectroscopy. The Rietveld plots represent a good structure fit between observed and calculated intensity with satisfactory R-factors. The bond distances Zr-O, P-O & Na-O are in agreement with their corresponding values for respective oxides. It was observed up to ~2.67 mol % (7.16 wt%) of strontium and ~3.56 mol % (14.46 wt%) of cesium could be simultaneously loaded into NZP formulations without significant changes of the three-dimensional framework structure

Substituted calcium zirconium phosphates (CZP)

Calcium Zirconium phosphate has been widely studied with a view to their potential application as a catalyst, ion exchanger and ion conductor. The structure of Molybdenum substituted calcium zirconium phosphate (CZP) was determined on the basis of crystal data of solid solutions. It was found that up to ~1.74 mol% (5.81 wt%) **of** Molybdenum could be loaded into CZP formulations without significant changes of the three-dimensional framework structure. The crystal chemistry of $Ca_{1-2x}Zr_4Mo_{2x}P_{6-2x}O_{24}$ (x=0.1-0.5) phases has been investigated using General Structure Analysis System (GSAS) programming. The Mo substituted CZP phases crystallize in the space group R-3 and Z=6. Powder diffraction data have been subjected to Rietveld refinement to arrive at a satisfactory structural convergence of R-factors. The PO_4 stretching and bending vibrations in the Infra red (IR) region have been assigned. SEM and EDAX analysis provide evidence of Mo in the matrix.

Perovskites

Substituted CaTiO$_3$

Ceramic phases Ca$_{1-x}$Sm$_x$TiO$_3$ (x=0.1-0.5) crystallize in the orthorhombic symmetry with space group Pbnm. In Samarium substituted calcium titanate, the perovskite matrix can be modified by partial substitution of up to a maximum limit of 8 mol% (33.40 wt%) of calcium by Samarium to yield a single phase polycrystalline solid solution beyond which extra Samarium is precipitated out of the matrix as an additional solid phase of Sm$_2$O$_3$. The substituted calcium titanate ceramic material has perovskite chains of distorted TiO$_6$ polyhedron interlinked through Ca atoms. The results show that the distortion of TiO$_6$ reduces the coordination number of Ca atoms to 8 resulting into eight acentric bonds in contrast to twelve for ideal cubic perovskite. The particle size along prominent reflections lies in the range of 11.23-105.4 nm for Ca$_{1-x}$Sm$_x$TiO$_3$ (x=0.1-0.5) ceramic materials. The crystallographic data and the structure model of Ca$_{1-x}$Sm$_x$TiO$_3$ (x=0.1-0.4) ceramic phase close to composition CaTiO$_3$ could be useful in the study of structure-property relationship of Samarium loaded nuclear waste forms. The atomic ratios of Ca, Ti and Sm are agreeable with the EDX analysis. The dielectric study of Ca$_{1-x}$Sm$_x$TiO$_3$ (x= 0.1 to 0.5) shows a diffuse ferro – para electric phase transition and relaxor behavior of the material. Frequency and temperature dependent dielectric constants and loss factor have been calculated each case.

Summary

Materials synthesis and characterization with a view to tailoring them for specific applications has the potential to assuage the enormous challenges besetting our socio-economic lives and resolve our energy crises. Materials science is an applied science concerned with the relationship between the structure and properties of materials. Atomic structure and chemical composition were once major focuses of materials science research. However, over the last few decades, this focus has changed dramatically as analytical chemistry, the electron microscope, X-ray diffraction, and a host of spectrometers have been developed that can analyze materials with accuracy. Materials scientists are facing unprecedented challenges in many areas, such as energy conversion and storage, microelectronics, telecommunication, display technologies, catalysis, and structural materials. Experimental methods generate increasing amounts of data. New computational methods, high-performance computer hardware, and powerful software environments are evolving rapidly. As a result, the importance of computational materials science is growing.

Ceramics are generally compounds between metallic and nonmetallic elements and include such compounds as oxides, nitrides, and carbides. Typically they are insulating and resistant to high temperatures and harsh environments. Many materials scientists are engaged in developing new ceramic materials and improving existing ones. This involves understanding the properties of the materials so as to tailor them for new applications. For example, doping--adding chemical elements that are not present in the starting material can change the electrical conductivity of a ceramic by many orders of magnitude, or change the type of conduction from electronic to ionic. Such conductivity changes are critical for electrochemical sensors and for materials used in clean energy production in solid-oxide fuel cells. Materials scientists also play a vital role in devising processes for joining ceramics with other materials, including metals and semiconductors, to make new products. Knowledge of properties is needed to overcome difficulties such as mismatches in thermal expansion that can become critical when temperature changes during the manufacture or use of a device. Products made entirely or partly from ceramics are everywhere around us today and promise to be prominent in the technologies of the future. The future of ceramic materials is even more interesting. Scientists have created ceramics that, while not as tough as metals, are many times tougher than those made just a few years ago. These tough materials are being used increasingly as parts in automobile engines because of their lightness and resistance to wear. Other ceramics have been made electrically conductive or able to allow oxygen ions to penetrate them. Both of these characteristics are needed for high-temperature fuel cells that can convert fuels such as natural gas directly to electricity more efficiently then any other method. Ceramics are being formed by methods similar to those used for mass production of plastic parts, so that increasingly intricate parts can be made cheaply. All these developments combined ensure that ceramics will continue to play important roles in modern life.

One of the recent applications of titania and zirconia based ceramic precursors is in immobilization and solidification of radioactive isotopes occurring in waste effluents coming out of nuclear establishments and power stations. Due to long term stability and integrity of the ceramic waste forms of high level nuclear waste, several countries have switched over from glass technology to ceramic technology of radwaste management. These and many more applications make ceramic material an interesting area of research and engineering sciences. It would be, therefore, desirable to study various routes of synthesis of titania and zirconia based ceramic materials. There are several solid state reactions, which take place on vitrification of high level nuclear waste. In order to develop and appreciate the "structure-property relationship" of various post vitrified ceramic phases, it has become logical and interesting to simulate the solid state reactions which might take place in process of conversion of nuclear effluents into the corresponding ceramic waste forms. The present work of the thesis has been divided in to four chapters:

Chapter I has been devoted to the introduction of the problem and survey of literature relevant to the proposed work. It contains brief the overview and scope of the work embodied in the thesis on titania and zirconia based ceramic materials used in management of nuclear waste. An exhaustive review of the literature has been given in this chapter describing critically various aspects of the problem. The crystallochemical behavior of the following group of materials has been investigated:

- Sodium zirconium phosphates
- Calcium zirconium phosphates
- Perovskites

The crystalline compounds of sodium zirconium phosphate $NaZr_2(PO_4)_3$, commonly known as NZP have been widely investigated since their discovery and continue to be of great interest as they are proposed in a variety of industrial and scientific applications. Phosphates with NZP ($NaZr_2P_3O_{12}$) structure form a broad family exhibiting a variety of physical properties. Their properties range from ionic conduction and low thermal expansion to radioactive nuclide immobilization. NZP is characterized by a three-dimensional framework which incorporates numerous cations in all the available sites. The stability and flexibility of the NZP structure allow Partial and complete isovalent, heterovalent and coupling substitution at all non-oxygen crystallographic sites and give rise to a large number of compounds with identical topology of connections between their structural units. The NZP compounds are lately receiving attention for their potential to be used in high anti-thermal shock applications, automobile industry, space, telescope technology, etc. Also, they can be used as catalyst supports, fast ionic conductors, hosts for immobilizing radioactive waste etc. A very important feature of those materials is the possibility to be designed with controlled thermal expansion coefficient. Additionally, some NZP members present high mechanical strength, good chemical stability, high melting point, great hardness and radiation resistance. In some cases the crystal lattice presents a highly anisotropic behavior. When anisotropy exists, it has a detrimental effect on mechanical strength because it causes micro stresses and consequently micro cracks.

During last few years a new structural family of low thermal expansion materials namely sodium zirconium phosphate (NZP) and calcium zirconium phosphate (CZP) has been developed. The group is characterized by a flexible framework structure belonging to the rhombohedral system with possibility of isomorphic replacements of various groups of elements. In recent years these solid solutions are receiving attention for their potential to be used as (i) ionic conductors and (ii) host material for radioactive waste immobilization because of their structural flexibility with respect to isomorphic ionic replacements and high stability against leaching reactions. Compounds with NZP skeleton are anisotropic, changing their dimension in apposite magnitudes when the counter ion of the skeleton is substituted or thermally affected. This fact is the basis of a series of materials with very low thermal expansion ($\alpha \sim 10^{-7}\,°C^{-1}$). It is this feature of the NZP and CZP skeleton which has generated interest in the study of mobility and resultant properties of these compounds. A model has been developed in order to put into simple parameters the atomic contributions and topological relations to the lattice dimensions. The structure of NZP compound in most cases belongs to the R-3c group whereas structure of CZP, Mo modified CZP fits into R-3.

Perovskite type oxides of general formula ABO_3 are important in material sciences, physics and earth sciences, for their electric properties and as a dominant mineral in the Earth's lower mantle. They are also well known for their phase transitions, which strongly affect their physical and chemical properties. Calcium titanate ($CaTiO_3$) was discovered for the first time in mineral form in 1839 by Gustav Rose, a German mineralogist. As ceramic material, $CaTiO_3$ is a key component of Synroc (type of synthetic rock used to store nuclear waste).The main advantage of the perovskite based ceramics as the matrix for the encapsulation of hazardous elements consists in high chemical durability and high absorption ability for heavy elements e.g. trans uranium elements that compose high level radioactive waste (HLW). The safe final disposal of these elements that are produced as waste from nuclear facilities has to fulfill many requirements. One of the requirements is to ensure high chemical durability as well as radiation and thermal stability of the ceramics forms where the radioactive nuclides were encapsulated or immobilized.

Chapter II deals with the basic principles and salient feature of the instrumental techniques used in the present work for the characterization of the synthetic phases used as a material for radwaste management and various electronic applications. These measurements have been carried out with the specific aim of gathering instrumental evidence in support of the authenticity of the synthetic phases with respect to their crystal structure and properties. The methods include powder X-ray diffraction, SEM, EDAX, impedance spectroscopy and Infra red spectroscopy. Depending on the intricacies within a program of research, or the requirements of an industrial process, a materials researcher will require access to a variety of scientific techniques. Each technique will have varying degrees of usefulness and importance depending on the research's requirements. Each technique will require a suitable knowledge, learning and expertise to operate at dexterity commensurate with the difficulty of the scientific objectives. Each technique has the potential to provide useful pathways to help reach a successful conclusion. A major emphasis of materials science is in understanding the elemental compositions and corresponding atomic structures present in materials of interest. This knowledge confirms a material's purity and suitability for use, and allows explanation for its properties and performance. Since a powder diffraction pattern is a set of peaks, some overlapped, superimposed on a smooth and slowly varying background, a Rietveld refinement can be thought of as a very complex curve fitting problem. The Rietveld method is a complex minimization procedure. It is not an active tool for crystal structure analysis. It can only slightly modify a preconceived model built on external previous knowledge. The starting parameters for such a model must be reasonable close to the final values. More over, the sequence into which the different parameters are being refined needs to be carefully studied.

In the present study the computer software General Structure Analysis System (GSAS) program with the EXPGUI graphical user interface was used for Rietveld analysis of the X-ray powder diffraction data. High quality powder diffraction data in combination with the Rietveld method allows refinement of a structural model (atomic coordinates, site occupancies and atomic displacement parameters) as well as profile parameters (lattice constants, peak shape, sample height, instrument parameters and background). X-ray powder diffraction (XRD) is a rapid analytical technique primarily used for phase identification of a crystalline material and can provide information on unit cell dimensions.

X-ray powder diffraction is a powerful non-destructive testing method for determining a range of physical and chemical characteristics of materials. It is widely used in all fields of science and technology. The applications include phase analysis, i.e. the type and quantities of phases present in the sample, the crystallographic unit cell and crystal structure, crystallographic texture, crystalline size, macro-stress and micro strain, and also electron radial distribution functions. Due to the importance and impact on science and technology X-ray diffraction has long been used to determine the atomic scale structure of materials. The technique is based on the fact that the wavelength of X-rays is comparable in size to the distances between atoms in condensed matter. Thus, when a bulk material that exhibits a long-range, periodic atomic order, such as a crystal, is irradiated with X-rays, it acts as an extended, almost perfect grating and produces a diffraction pattern showing numerous sharp spots, called Bragg diffraction peaks. By measuring and analyzing the positions and intensities of Bragg peaks, it is possible to determine the spatial characteristics of the grating, i.e. to determine the three-dimensional (3D) atomic arrangement in bulk crystals. This is the essence of the so-called "crystal structure" determination by X-ray diffraction (XRD).

Every material has a unique set of electrical characteristics that are dependent on its dielectric properties. Accurate measurements of these properties can provide scientists and engineers with valuable information to properly incorporate the material into its intended application for more solid designs or to monitor a manufacturing process for improved quality control. Impedance Spectroscopy, measures the dielectric properties of a medium as a function of frequency. It is based on the interaction of an external field with the electric dipole moment of the sample, often expressed by permittivity. It is also an experimental method of characterizing electrochemical systems. This technique measures the impedance of a system over a range of frequencies, and therefore the frequency response of the system, including the energy storage and dissipation properties, is revealed. A dielectric materials measurement can provide critical design parameter information for many electronics applications. For example, the loss of a cable insulator, the impedance of a substrate, or the frequency of a dielectric resonator can be related to its dielectric properties. The information is also useful for improving ferrite, absorber, and packaging designs. More recent applications in the area of industrial microwave processing of food, rubber, plastic and ceramics have also been found to benefit from knowledge of dielectric properties.

The dielectric constant and loss are important properties of interest to electrical engineers because these two parameters, among others, decide the suitability of a material for a given application. Dielectric relaxation is studied to reduce energy losses in materials used in practically important areas of insulation and mechanical strength. An analysis of build up of polarization leads to the important Debye equations. The Debye relaxation phenomenon is compared with other relaxation functions due to Cole-Cole, Davidson-Cole and Havriliak-Negami relaxation theories. The behavior of a dielectric in alternating fields is examined by the approach of equivalent circuits which visualizes the lossy dielectric as equivalent to an ideal dielectric in series or in parallel with a resistance. Finally the behavior of a non-polar dielectric exhibiting electronic polarizability only is considered at optical frequencies for the case of no damping and then the theory improved by considering the damping of electron motion by the medium.

Chapter III deals with the synthesis procedure and experimental data obtained on selected titania, zirconia, phosphate and samarium based ceramic materials. The compositional details of the materials and their syntheses route have been presented in this chapter. The following ceramic phases have been synthesized during the course of investigations:

Sodium zirconium phosphates (NZP):

- Sodium Zirconium Metal Phosphate $Na_{1-x}M_{x/2}Zr_2P_3O_{12}$ [M = Sr and x = 0.1-1.0]
- Co- substitution in Sodium Zirconium Metal Phosphate $Na_{1-x}(M'_{1.33}M''_1)_xZr_2P_3O_{12}$ [M'=Cs, M''= Sr and x = 0.1-1.0]

Calcium zirconium phosphates (CZP):

- Calcium Zirconium Metal Phosphate $Ca_{1-2x}Zr_4M_{2x}P_{6-2x}O_{24}$ [M = Mo and x = 0.1-0.5]

Perovskites:

- Metal Calcium Titanate $Ca_{1-x}M_xTiO_3$ [M = Sm and x = 0.1-0.5]

The samples have been prepared by solid state sintering which involves repeat steps of mixing, grinding and heating of oxides until a single-phase and high density polycrystalline material is obtained. The objective of ceramic route synthesis of titania and zirconia based ceramic precursors is to understand the crystallochemical interaction of various cations with the host matrix and their effect on crystallographic and electrical properties of materials. The materials have been characterized by powder X-ray diffraction and other analytical methods such as scanning electron microscopy, energy dispersive X-ray analysis and infrared spectroscopy etc. Their morphology and micro structure examined by SEM and EDAX analysis. The temperature and frequency dependence of capacitance and impedance of samarium substituted calcium titanate were measured by means of the conventional four-point probe technique. For detailed investigation on electrical behavior, dielectric measurements are made over the range of frequencies covering audio and radio frequency regions. The results are plotted as frequency vs. dielectric loss factor (tan delta), temperature vs loss factor, frequency Vs dielectric constant and temperature vs dielectric constant.

Chapter IV summarizes the results and discussion of the experimental observations. X-ray powder diffraction data for the Rietveld refinement of various phases were collected and refinement was performed on GSAS software. Rietveld refinement provides a complete structural model and a fit of entire diffractogram covering all prominent reflections including line positions, line widths and relative line intensities. A Rietveld model employs lattice parameters, atomic positions, occupation factors, thermal parameters, interatomic distances and bond angle data etc. The structure model in each case has been suggested with the help of graphics software's like PLATON, Ball & Stick and ORTEP etc. Particle size calculation using Scherrer's formula suggest that crystallite size along prominent reflections belongs to nano range in most of the synthetic phases.

Substituted sodium zirconium phosphate (NZP)

Refinement of powder X-ray diffraction data of NZP related phases show that the solid solutions of substituted sodium zirconium phosphates crystallize in the rhombohedral system with space group R-3c. Reprocessing of spent nuclear power reactor fuel involves various steps of immobilization and final disposal of precipitate forming elements like Zr, Cs and Sr that can not be concentrated by evaporation of liquid waste forms. It is therefore, desirable to understand crystallochemical fixation of Cs and Sr atoms into NZP like ceramic matrix for practical applications. The structural analogue of sodium zirconium phosphate is a potential material for immobilization of cesium and strontium from radio active waste along with other radio nuclides. The existence of Cs, Sr and several other radionuclide containing NZP-type structures was assessed on the basis of crystal chemical principles. In this context, two major types of formulations of simulated NZP waste forms have been synthesized whose chemical compositions are as follows: $Na_{1-x}Sr_{x/2}Zr_2P_3O_{12}$ (x=0.1-1.0) and $Na_{1-x}(Cs_{1.33}Sr_1)_xZr_2P_3O_{12}$ (x = 0.1 to 1.0). It is in this background that the complete crystallographic characterization of such simulated waste-forms becomes interesting for process engineering and system development for immobilization of light water reactor's low level waste effluents in NZP matrix. The powder XRD data showed that a solid solutions of $Na_{1-x}Sr_{x/2}Zr_2P_3O_{12}$ (x=0.1-1.0) and $Na_{1-x}(Cs_{1.33}Sr_1)_xZr_2P_3O_{12}$ (x = 0.1 -1.0) WO_x loaded waste forms are formed which are isostructural to $NaZr_2(PO_4)_3$.

Solid solution of $Na_{1-x}Sr_{x/2}Zr_2P_3O_{12}$ (x=0.1-1.0) were synthesized by conventional solid state reaction method. The material has been characterized by XRD, SEM-EDAX and FTIR analysis. The SrNZP phases crystallize in the space group R-3c and Z = 6. Powder diffraction data have been subjected to Rietveld refinement to arrive at a satisfactory structural convergence of R-factors. The PO_4 stretching and bending vibrations in the infrared region have been assigned. It was observed up to ~1.98 mol% (6.07 wt%) of strontium could be loaded into NZP formulations without significant changes of the three-dimensional framework structure.

The crystal structure of simultaneously substituted cation (viz-Cs & Sr) in NZP matrix has been also investigated and compared them with those of the substrate matrix. Refinement of powder X-ray diffraction data of

$Na_{1-x}(Cs_{1.33}Sr_1)_xZr_2P_3O_{12}$ (x = 0.1-1.0) shows that the Cs and Sr substituted NZP crystallizes in the rhombohedral (R-3c space group) structure. Crystal data and structural parameters of the material have been refined to a satisfactory convergence with reasonable values residual factors (R_p & R_{wp}). The calculated values of P-O and Zr-O bond lengths and O-M-O bond angles are in good agreement with the expected values. The material has been characterized by XRD, SEM-EDAX and FTIR analysis. It was observed from the crystallographic study of $Na_{1-x}(Cs_{1.33}Sr_1)_xZr_2P_3O_{12}$ (x = 0.1 to 1.0) up to ~2.67 mol % (7.16 wt%) of strontium and ~3.56 mol% (14.46 wt%) of cesium could be simultaneously loaded into NZP formulations without significant changes of the three-dimensional framework structure, beyond these limits traces of minor secondary phase of cesium strontium zirconium phosphate starts appearing along with the solid solution.

Substituted calcium zirconium phosphate (CZP)

The structure of Molybdenum substituted calcium zirconium phosphate (CZP) was determined on the basis of crystal data of solid solutions. The crystal chemistry of mono phases of composition $Ca_{1-2x}Zr_4Mo_{2x}P_{6-2x}O_{24}$ (x=0.1-0.5) has been investigated using General Structure Analysis System (GSAS) programming. The Mo substituted CZP phases crystallize in the space group R-3 and Z=6. Powder diffraction data have been subjected to Rietveld refinement to arrive at a satisfactory structural convergence of R-factors. The unit cell volume and polyhedral (ZrO_6 and PO_4) distortion increases with rise in the size and mole % of loaded cation in the CZP matrix. The presence of orthophosphate anions in the crystal structure was confirmed with the IR spectroscopy. The absorption bands in the range between 1250-1022 cm^{-1} and 650-507 cm^{-1} are assigned to stretching and bending vibrations of P-O bonds of the PO_4 tetrahedron respectively. The stretching vibrations occur between 1270-1020 cm^{-1} as υ_3 band, the symmetric stretching υ_1 and anti symmetric bending υ_4 vibrations are observed in the regions 990-900 cm^{-1} and 640-505 cm^{-1} respectively. SEM and EDAX analysis provide evidence of Mo in the matrix. It was found that upto ~1.74 mol% (5.81 wt%) of molybdenum could be loaded into CZP formulations without significant changes of the three-dimensional framework structure.

Substituted calcium titanate

Perovskite compounds are favorable functional materials since they accommodate both large (A-site) and small (B-site) cations. In this structure, it is possible to incorporate cations of different sizes due to distortion of the ideal cubic structure. $CaTiO_3$ is a perovskite oxide material that has been widely used in electronic devices as dielectric material.

The X-ray diffraction data shows that the matrix of $CaTiO_3$ can be modified by partial substitution of calcium by samarium to yield a single phase polycrystalline solid solution. Perovskite phases of $Ca_{1-x}Sm_xTiO_3$ (x = 0.1 -0.5) crystallize in orthorhombic symmetry with space group Pbnm. Rietveld analysis of the above phases show the agreement in the expected and calculated R factors and goodness of fit and yields acceptable reliability factors: Rp, Rwp and RF^2. The Ti site of Samarium substituted calcium titanate maintains six co-ordinations with the shortest Ti(5)-O(1) bond of 1.38192 Å and the longest one Ti(5)-O(1) of 2.46815 Å. The distortion of TiO_6 reduces the coordination number of Ca/Sm atoms to 8 resulting into eight acentric bonds in contrast to twelve for ideal cubic perovskite. The interatomic distances between Ti(5) and the apex oxygen atoms of the octahedron have been found to be 1.92311 Å whereas the two sets of planar oxygen bond distances Ti(5)-O(1) are 1.38192 and 2.46815 Å respectively. Dielectric properties as a function of temperature have been studied for all phases of $Ca_{1-x}Sm_xTiO_3$ (x = 0.1-0.5) which shows that the samarium doped calcium titanate exhibits almost temperature independent dielectric constant, which is about 3-4 times greater than the corresponding value of unsubstituted calcium titanate. In Samarium substituted calcium titanate, the perovskite matrix can be modified by partial substitution of up to a maximum limit of ~8 mol% (33.40 wt%) of calcium by Samarium to yield a single phase polycrystalline solid solution beyond which extra Samarium is precipitated out of the matrix as an additional solid phase of Sm_2O_3.

Dielectric properties as a function of frequency and temperature have been studied for the phases $Ca_{1-x}Sm_xTiO_3$ (x = 0.1-0.5). The frequency dependence of dielectric constant and loss at Different temperature for the Calcium Titanate (CT) series modified by the substitution of Sm. Dielectric constant & loss factor (ε' and $\tan\delta$) shows decreasing trend with increase in frequency. The dielectric constant of $Ca_{1-x}Sm_xTiO_3$ (x= 0.1 to 0.5) at lower frequencies is higher than the corresponding values in higher frequency region. This is the normal behavior of ferroelectric materials.

It was observed that the value of dielectric constant of each specimen at higher frequency gets markedly dropped. This phenomenon can be explained in term of interfacial polarization. This built up of charges at the grain-grain boundary interface is responsible for large polarization, therefore high dielectric constant at low frequency.

The fall in dielectric constant arises from the fact that polarization does not occur instantaneously with the application of the electric field because of inertia. The delay in response towards the impressed alternating electric field leads to loss and decline in dielectric constant. At low frequencies, all types of polarization such as interfacial, atomic, dipolar, ionic and electronic contribute. As frequency is increased, those with large relaxation times cease to respond and hence the decrease in dielectric constant. The dependence of dielectric constant and loss factor at different temperature was investigated. As typical of normal ferroelectrics, dielectric constant increases gradually with increment in the temperature due to interfacial polarization becoming more dominant as compared to the dipolar polarization and passes through a maximum (Curie temperature, T_c) and then decreases due to the phase transition from ferroelectric to the paraelectric phase. The temperature dependent dielectric loss(tanδ) slowly increase with increase of temperature at all frequencies and temperatures up to 300°C. The variation of tanδ at low temperature is smaller compare to that at higher temperature. At higher temperature conductivity begin to dominate resulting in an increase of tanδ, and hence at higher temperature tanδ is typically associated with the loss by conduction.

The dielectric relaxation phenomenon in ferroelectric materials reflects the delay (time dependence) in the frequency response of a group of dipoles when submitted to an external applied field. When an alternating voltage is applied to a sample, the dipoles responsible for the polarization are no longer able to follow the oscillations of the electric field at certain frequencies. The field reversal and the dipole reorientation become out-of-phase giving rise to a dissipation of energy. Over a wide frequency range, different types of polarization cause several dispersion regions and the critical frequency, characteristic of each contributing mechanism, depends on the nature of the dipoles.

The dielectric study of $Ca_{1-x}Sm_xTiO_3$ (x= 0.1 to 0.5 shows a diffuse ferro – paraelectric phase transition and relaxor behavior of the material.